DARK SIDE OF THE MOON

GERARD J. DEGROOT

DARK SIDE OF THE MOON

The Magnificent Madness
of the American Lunar Quest

New York University Press • *New York and London*

NEW YORK UNIVERSITY PRESS
New York and London
www.nyupress.org

Library of Congress Cataloging-in-Publication Data
De Groot, Gerard J., 1955–
Dark side of the moon : the magnificent madness of the American lunar quest /
Gerard J. DeGroot.
p. cm.
Includes bibliographical references and index.
ISBN–13: 978–0–8147–1995–4 (cloth : alk. paper)
ISBN–10: 0–8147–1995–3 (cloth : alk. paper)
1. Project Apollo (U.S.)—History.   2. Space flight to the moon—History—20th century.
3. Moon—Exploration—20th century.   4. United States—History—1953–1961.
5. United States—History—1961–1969.   I. Title.
TL789.8.U6A5318   2006
629.45'4—dc22   2006016116

Manufactured in the United States of America
10 9 8 7 6 5 4 3 2 1

In memory of my father,

Jan DeGroot,

1920–1987,

a great craftsman who kept us fed by building rockets

Contents

Acknowledgments   ix

Preface   xi

1   Fly Me to the Moon   1

2   Slaves to a Dream   12

3   What Are We Waiting For?   29

4   Sputnik   45

5   The Red Rocket's Glare   61

6   Muttnik   79

7   Rocket Jocks   100

8   Before This Decade Is Out   121

9   The Sleep of Reason Produces Monsters   153

10 Lost in Space   183

11 Sacrifices on the Altar of St. John   205

12 Merry Christmas from the Moon   223

13 Magnificent Desolation   233

14 Nothing Left to Do   255

Notes   271

Bibliography   289

Index   293

About the Author   321

Acknowledgments

I would like to acknowledge assistance given by the staff at the NASA History Office, the Kennedy Space Center, the Johnson Space Center, the Lyndon Baines Johnson and John F. Kennedy Libraries, and the National Air and Space Museum in Washington, D.C. Financial assistance was generously provided by the Carnegie Trust for the Universities of Scotland and by the University of St. Andrews. Friends and colleagues who provided small favors have been thanked in my own personal way, but I would like to give special thanks to David Onkst, who did sterling work in the NASA archives.

I must also thank my family—Sharon, Natalie, and Josh—who bravely endured my stories about the Moon.

Preface

A funny thing happened on my way to Tranquillity Base. I suppose I could call it a revelation. This book, provisionally titled *One Giant Leap*, was originally intended as an antidote to my last book, *The Bomb: A Life*. Writing about nuclear weapons left me depressed, cynical, forlorn, and scared. After that experience, I craved something hopeful and uplifting and therefore looked to my childhood for a good wholesome story to retell. I decided to write a book about the heroes of my youth—the astronauts who took America to the Moon. I suppose that's self-indulgent, but I didn't care. I'd earned it after doing time with the Bomb.

When I started looking at the lunar program, I found heroes aplenty. But I also found a gang of cynics, manipulators, demagogues, tyrants, and even a few criminals. I discovered scheming politicians who amassed enormous power by playing on the public fascination for space and the fear of what the Russians might do there. Quite a few people got rich from the lunar mission; some got very rich indeed. The Moon mission was sold as a race that America could not afford to lose— a struggle for survival. Landing on the Moon, it was argued, would bring enormous benefit to all mankind. It would be good for the economy, for politics, and for the soul. It would, some argued, even end war.

Referring to the shallow nature of the *Apollo 11* coverage on television and in newspapers, Edwin Diamond, a senior editor at *Newsweek*, wrote, a short time after the launch:

> Little of the flesh and blood vitality—and human frailties—of the past decade of the American space venture were offered. . . . Among the missing stories, to take only the most obvious examples, were the Cold War beginnings of the space program; John F. Kennedy's search for a space spectacular "that the U.S. could win"; the spurious nature of the "Moon race" with the Russians (we raced only ourselves); the separate fiefdoms and the abrasive clash of personalities in NASA; the logrolling politics of space appropriations and decisions that put the

Manned Spacecraft Center in Texas and other installations in Louisiana and Massachusetts; the shoddy workmanship of some of the biggest U.S. firms and the slipshod Government management procedures that led to the death of three astronauts—in short, the full, as opposed to the official story of Apollo.[1]

This book provides those missing stories and some others Diamond did not know about back in 1969. My aim has been to cut through the myths carefully constructed by the Kennedy and Johnson administrations and sustained by NASA ever since.

The popularity of space stories is perhaps understandable. We all love heroes, we all enjoy a great adventure. While we no longer travel to the Moon, we still travel nostalgically to that era when the Moon made us feel good. For many of us who remember the sixties, Apollo represents a safe harbor in a sea of cynicism, violence, and despair.

Myths are, however, dangerous, especially when they're used to manipulate contemporary opinion. NASA still cashes in on the glories of forty years ago. The problems of the space program today, for instance that of the underperforming and sometimes dangerous Space Shuttle, can be traced back to decisions made in the 1960s, in particular the misguided emphasis upon manned space travel. Putting men in space was an immensely expensive distraction of little scientific or cultural worth. The American people, in other words, were fleeced: they were persuaded to spend $35 billion on an ego trip to the Moon, and then were told that a short step on the desolate lunar landscape was a giant leap for mankind.*

The ground rules of the space race were set the moment the Russians stuffed a poor little dog into a space capsule on November 3, 1957. From that point forward, only those achievements carried out by living, breathing things really mattered. The Russians might have set the terms of the race, but the Americans gladly went along, because they understood that the contest would only capture the public imagination if it were turned into a human adventure. The early space pioneers, Wern-

---

* Estimating the cost of the mission to the Moon is nearly impossible given the difficulty in deciding what to count. Simply adding up NASA yearly budgets would be wrong, since not all the money went toward the task of a Moon landing. The figure of $35 billion is at the top end of estimates but still widely accepted by most analysts. The sharp-eyed reader will notice different figures quoted at various points in this book, as a result of differences of opinion among those being quoted.

her von Braun foremost among them, played to this egotism. He realized that the only way to get money to fund his dreams was if space had a face. No bucks without Buck Rogers.*

Exploration could have been carried out by robots, but robots could never be heroes. Yet putting men into the capsules added huge complexities to the space equation. It meant that every voyage had to be a round trip, and that sophisticated life support systems had to be added to cater to the astronaut's bodily needs. The capsules had to be not just laboratories, but also toilets, kitchens, and bedrooms. The air had to be pure, the water clean, and the climate warm. All this severely limited what could be explored. Robots can easily travel to the far reaches of the solar system, while men can't make it beyond Mars and certainly can't land on Venus.

Expressed in the terms set by the Soviets and the Americans, the lunar race was shallow and trivial. The two superpowers behaved like two bald men fighting over a comb. The Moon became the target, not because it was important, but because it was there. It was far away, threatening and mysterious, but still close and familiar enough to make a journey possible. By mutual agreement, it became the finish line in the space race. What was important was not what the first explorers might find, but rather that they were there at all. The limited nature of this endeavor explains why the American space program has been caught in a state of purposeless wandering ever since Neil Armstrong set foot on the Moon. No one knew what to do next, because the only goal that had ever mattered had been achieved.

When Armstrong and Buzz Aldrin landed on the Moon, the Soviet premier Leonid Brezhnev congratulated the brave astronauts who, he said, had overcome the limitations of their machinery. Americans at the time thought the comment churlish but, in retrospect, it seems remarkably perceptive. Scientists and engineers stretched available technology as far as it could possibly go, thus allowing the mission to proceed much earlier than logic or caution should have allowed. Lives were risked—nearly every mission came close to catastrophe. Going to the Moon was not just a supreme technological achievement, it was also a magnificent artistic endeavor, requiring huge reserves of imagination, faith, and bravery. But all that sublime effort was devoted toward a stunt that had no real purpose other than to kick lunar dust in Soviet

---

* A phrase used many times, including by Tom Wolfe.

faces. The mission was a brilliant deception, a glorious swindle—"magnificent desolation."*

Those who look to history for heroes will find this book disappointing, but I make no apologies for this exclusion. From the glory days of Mercury to the present sad era of the Space Shuttle, NASA has cashed in on the American public's willingness to indulge a fantasy of manned space exploration. Space agency self-interest has been marketed as a national good, and at times as the greatest hope for mankind. It is time that we shield ourselves from the blinding light of the Moon, and look instead at its dark side.

* The first words uttered by Aldrin when he stepped on the Moon.

Fly Me to the Moon

Lagari Hasan Celebi is probably the first person to have flown. He built a rocket, packed it with 25 kilos of gunpowder and put on top a conical wire cage, into which he climbed. A friend then lit the fuse and a great explosion threw Celebi from his launch pad on the banks of the Bosphorus 300 meters into the air. At his apogee, he opened some homemade wings and drifted safely back to Earth. Or so it is claimed.

The year was 1623. Celebi performed the stunt to impress Sultan Murad IV on the occasion of his daughter Kaya's birth. "Your Majesty, I leave you in this world while I am going to have a talk with the prophet Jesus," he told the sultan before boarding his rocket. "Your Majesty, Prophet Jesus sends his greetings to you,"[1] The crowds went wild. So, too, did the sultan, who awarded Celebi a pouch of gold and a commission in the cavalry.

On that first manned rocket mission, two important principles of space travel were established. The first was that the heavens are always more impressive in the imagination than in reality. The second was that exploits in outer space are important primarily for what they achieve back on Earth.

In 1869, Edward Everett Hale, a Unitarian minister, wrote an article for the *Atlantic Monthly* entitled "The Brick Moon." In the story, a group of idealistic students devise a plan to build a satellite to serve as a navigational aid to those back on Earth. As the title implies, their satellite was constructed entirely of brick, in order to withstand the heat of passing through the atmosphere. It was propelled into orbit by being rolled down a huge groove cut on a mountainside. At the end of the groove were two water-powered flywheels that flung it into the heavens.

Once in orbit, the Brick Moon exceeded wildest expectations. Not only did it help stranded sailors and explorers, it also provided an example of a harmonious community—the kind of peace and goodwill missing back on Earth.

In a tropical climate they were forming their own soil, developing
their own palms, and eventually even their bread-fruit and bananas,
planting their own oats and maize, and developing rice, wheat, and all
other cereals, harvesting these six, eight, or ten times . . . in one of our
years. . . . For them, their responsibilities are reduced in somewhat the
same proportion as the gravitation which binds them down. . . . This
decrease of responsibility must make them as light-hearted as the loss
of gravitation makes them light-bodied.[2]

The thirty-seven inhabitants lived in a self-sufficient outer space
Utopia. President Ulysses Grant decided that the Brick Moon was "the
biggest thing since Creation, save for the invention of bourbon whisky
and the Havana cigar.[3] He was probably joking. One hundred years
later, Richard Nixon called the *Apollo 11* mission to the Moon "the greatest week in the history of the world since the Creation."[4] He, unfortunately, wasn't joking.

Aviation's early philosophers, men like Alfred Lawson, engineer
and editor of *Aircraft* magazine, thought that the age of flight would
usher in a "superior type of man . . . Alti-man—a superhuman who
will live in the upper strata of the atmosphere and never come down
to Earth at all." Lawson, of course, lived long before there were cheap
charter flights to Cancun, packed with people who enjoy a beer and
chaser for breakfast. His vivid imagination never quite reckoned on
air rage. He demonstrates rather well the weird effect that technology
has on human perception. Virtually every major technological development, including television, nuclear power, and the Internet, was
initially expected to improve man, either physically or morally or
both. Instead, they have rendered man more efficient in his moral cor-
ruption.

The "winged gospel," as historian Joseph Corn has called it, was
based on the idea that once man learned how to fly (even if in an aluminum tube), his soul would also take flight. Man's soaring soul would
enable him to figure out how to end poverty, disease, war, and all forms
of suffering. Only in the air, apparently, is real tolerance and charity
possible.[5] The idea somehow survived, even though airplanes were
soon used for ghastly purposes, over Guernica, Rotterdam and Hi-
roshima. The same fantasies can still be found in more recent works by
Dandridge Cole, Krafft Ehricke, and Gerard O'Neill—men who should

have known better. It also explains Neil Armstrong's "giant leap for mankind." Among space enthusiasts, progress is often confused with spatial movement.

Ehricke, who during the Second World War helped build the German V-2 rocket that rained down upon London and Antwerp, later became a sort of cosmic guru who philosophized about the civilizing potential of the cosmos. In space, he argued, "there would probably be no more need for government than in a modern hotel." In 1957, just before the launch of Sputnik, he laid down his Three Laws of Astronautics:

First Law: Nobody and nothing under natural laws of this universe impose any limitations on man except man himself.

Second Law: Not only the Earth, but the entire solar system, and as much of the universe as he can reach under the laws of nature, are man's rightful field of activity.

Third Law: By expanding through the universe, man fulfills his destiny as an element of life, endowed with the power of reason and the wisdom of the moral law within himself.[6]

Ehricke decided that "as we place the extraterrestrial domain into the service of all people, we may be permitted to hope for the greatest benefit of all: that the ugly, the bigoted, the hateful, the cheapness of opportunism and all else that is small, narrow, contemptible and repulsive becomes more apparent and far less tolerable from the vantage point of the stars than it ever was from the perspective of the mudhole."[7] In other words, true perfection would be found in the synthetic environment of a rotating aluminium can orbiting 500 miles out in space, where there's not enough water to make a mudhole.

Gravity, apparently, is a corrupt tyrant—a power that keeps man from realizing spiritual nobility. True liberation can only come in an environment of weightlessness. In 1929, Cambridge physicist J. D. Bernal prophesied that, as a result of the very distinct forces acting upon man in space and on Earth, man would eventually divide into two species— Earthkind and Spacekind, the latter being the higher evolutionary form. C. S. Lewis, however, wasn't so sure. In *Out of the Silent Planet*, he penned one of the best attacks on space age fantasists—or those who wish "to open a new chapter of misery for the universe":

It is the idea that humanity, having now sufficiently corrupted the planet where it arose, must at all costs contrive to seed over a larger area . . . beyond this lies the sweet poison of the false infinite—the wild dream that planet after planet, system after system, in the end galaxy after galaxy, can be forced to sustain, everywhere and forever, the sort of life which is contained in the loins of our own species—a dream begotten by the hatred of death upon the fear of true immortality, fondled in secret by thousands of ignorant men and hundreds who are not so ignorant.

Lewis believed that, in the end, the laws of physics would defeat man's ambition. The vast distances in space, he decided, were "God's quarantine regulations."[8]

Space, we need to remind ourselves, is a vacuum, a void, nothingness. It is an empty vessel filled by men's imaginations. The more dispiriting life on Earth has seemed, the more attractive the cosmos has become. Bernard Smith, who launched a liquid-fueled rocket in 1933, had difficulty finding paid employment of any sort during the Great Depression. As a rocket scientist, he felt that fulfillment was his birthright. But Earth proved stubbornly ungenerous. It was "a lousy planet," he complained, and "the rocket ship was the only way to get off of it."[9]

Smith would probably have enjoyed sharing a vodka with Konstantin Tsiolkovsky, the father of Russian rocketry, and a man who also felt the mystical pull of space.

The basic drive to reach out for the Sun, to shed the bonds of gravity, has been with me ever since my infancy. Anyway, I distinctly recall that my favorite dream in very early childhood, before I could even read books, was a dim consciousness of a realm devoid of gravity where one could move unhampered anywhere, freer than a bird in flight. What gave rise to these yearnings I cannot say . . . but I dimly perceived and longed after such a place unfettered by gravitation.[10]

"My interest in space travel was first aroused by the famous writer of fantasies Jules Verne," Tsiolkovsky confessed. "Curiosity was followed by serious thought."[11] In between his practical work on rocket engines, Tsiolkovsky mused at length about the world that could be created out in the cosmos. He wrote of a "Great Migration" in which a select group of human beings would leave behind imperfect Earth and create a Shangri La up there. Since Tsiolkovsky's space age utopia was essentially a socialist community, his ideas delighted Lenin. The Bolsheviks consequently named Tsiolkovsky "Father of Cosmonautics."

Tsiolkovsky saw weightlessness and liberty as one and the same. In outer space one could find freedom and fulfillment—"the perfection of mankind and its individual members."[12] Alexander Bogdonov, a contemporary, took Tsiolkovsky's dream one step further, imagining a communist utopia on, appropriately, the Red Planet. The community he described in his novel *Red Star* (1908) was free of injustice, bigotry, poverty, or suffering of any sort.

Modes of travel were easily imagined; the laws of physics easily ignored. Hale used spinning flywheels; Verne a huge cannon. In 1638, the English bishop Francis Godwin (aka Domingo Gonsales) wrote fancifully of being pulled aloft by a flock of wild swans. Cyrano de Bergerac's explorers were taken to the lunar surface by the force of evaporating dew, while the American writer Joseph Atterley simply imagined into being an antigravity substance called lunarium. Since reality did not impede upon methods of propulsion, it's no wonder that it did not get in the way of dreams of what would be found in outer space. As G. Edward Pendray, journalist and sometime science fiction author, admitted, "imaginations could outpace dull practical considerations with a velocity comparable to that of light."[13]

In truth, real rockets were both expansive and limiting: they encouraged space fantasists but also eventually injected a sense of reality into their dreams. Once man began to fire rockets into the sky he was forced to come to terms with how difficult space travel is, how great the distances are, and how little can be achieved.

Experiments conducted by the American Robert Hutchings Goddard demonstrated the difficulty of leaving Earth. Unlike Tsiolkovsky or Smith, he was never moved to embellish his technical computations with elaborate cosmic fantasies. Nor was he motivated by the lure of scientific discovery. Goddard was more practical; he was convinced that the eventual cooling of the Sun would force humans into outer space in order to insure the continuance of the race. Toward that end, he sent an article entitled "A Method of Reaching Extreme Altitudes" to the Smithsonian Institution in 1917. The Smithsonian responded with a grant of $5,000 to fund further experimentation. In the article, Goddard proposed that a rocket could be flown to the Moon and that its arrival there could be marked by a large explosion that would be seen back on Earth.*

Cynical journalists had great fun with America's rocket man, pointedly remarking that Verne, at least, admitted he was writing fiction. On January 13, 1920, the *New York Times* accused Goddard of not knowing "the relation of action to reaction, and of the need to have something better than a vacuum against which to react." The paper clearly didn't understand the basics of rocketry, nor indeed Newton's third law of motion, but not many people did back then. Goddard, who had a chip on both shoulders, took the rebuff personally, as he did virtually everything in life. He wanted to be celebrated as a visionary, rather like Tsiolkovsky, but most people thought him a lunatic. A master of self-pity, he went into a sulk lasting twenty-five years, in the process insuring that his ideas never received the attention they perhaps deserved.

For this reason, Americans missed the significance of an event on March 16, 1926, when Goddard launched the world's first liquid-propelled rocket from his aunt's farm near Auburn, Massachusetts. The rocket reached a speed of 60 miles per hour and a height of 41 feet. Chastised by the earlier reaction to his ideas, Goddard refused to publicize the event. Journalists nevertheless got wind of it and took delight in mentioning how his "Moon rockets" remained stubbornly on Earth. One paper pointed out that the rocket had missed its target by "238,799 1/2 miles."[14]

As Goddard's rockets grew, so too did the intolerance of his neighbors. Feeling ever more victimized, he took his ideas out to New Mexico, an area big enough to accommodate his reclusive temperament. Funding for the move, and for the establishment of a purpose-built facility, came from Daniel Guggenheim, who was fascinated by rockets. Goddard, hopeless at selling himself, relied on Charles Lindbergh for promotion. The facility, not far from a place appropriately called High Lonesome, was located in the Eden Valley, near Roswell, the town made famous when little green aliens supposedly visited in the 1950s.

Anne Morrow Lindbergh recalled a conversation between her husband and Goddard on the potential of rockets. "He had envisaged man's landing on the Moon and even travelling to the planets, but he was cautious and practical when talking about the next step. 'Theoretically,' he

---

* This idea would resurface in the 1950s when some American scientists looked seriously into the possibility of exploding an atom bomb on the Moon, as a publicity stunt.

said, 'it would be possible to design a multi-stage rocket capable of reaching the Moon.' But he figured it might cost a million dollars."[15]

In May 1940, Goddard, with the help of Guggenheim's son, Harry, secured a meeting with representatives from the American armed forces to see if he could interest them in rockets. They were unimpressed, however, remarking that the "next war will be won with the Trench mortar."[16] While this might seem one of the most shortsighted opinions of all time, American officers were in fact displaying impressive wisdom. At that time, Goddard's rockets offered at best an uncertain return from a huge investment. As the German experience in World War II would eventually reveal, the rocket was a weapon before its time.

Goddard's greatest fault was his paranoia. Trusting no one, he was extremely reluctant to share his ideas. Yet scientific research cannot thrive in a vacuum. Newton, remember, attributed his success to the fact that he had stood on the shoulders of giants. Goddard preferred to stand alone, shooting ever more powerful rockets into New Mexico's limitless sky. The mistakes he made, and they were many, were made in private, where they were less likely to get observed and corrected. Theodore von Kármán, the Hungarian-born scientist and engineer who led the Guggenheim Aeronautical Laboratory at the California Institute of Technology (GALCIT), was not impressed with Goddard's contribution to rocketry:

> There is no direct line from Goddard to [the] present-day. . . . He is on a branch that died. He was an inventive man and had a good scientific foundation, but he was not a creator of science, and he took himself too seriously. If he had taken others into his confidence, I think he would have developed workable high-altitude rockets and his achievements would have been greater than they were. But not listening to, or communicating with, other qualified people hindered his accomplishments.[17]

Goddard died on August 10, 1945, the day the Japanese surrendered. Just before his death he told a reporter: "I feel we are going to enter a new era. . . . It is just a matter of imagination how far we can go with rockets. I think it is fair to say you haven't seen anything yet."[18]

As Goddard discovered, neither the American government nor the military were very interested in space. The same could not be said for the American people. Quite a few were convinced that outer space held an assortment of adventures, delights, mysteries, and dangers, and that it behooved man to go out and embrace them. Those who didn't feel the spiritual pull of space still believed that the cosmos offered enormous potential for wealth. The imagined riches that once drove Cortez to Peru caused a later generation of would-be explorers to look skyward.

Space was interesting precisely because of the flourish that imaginative entrepreneurs gave it. At the 1901 Pan American Exposition in Buffalo, New York, the show business entrepreneur Frederick Thompson unveiled his carnival extravaganza, which he called "A Trip to the Moon." The public boarded a 30-foot spaceship called *Luna* which "flew" them to the Moon. After a bumpy landing they disembarked on an eerie landscape full of stalactites and crystals. A group of midget "Selenites" then escorted the visitors to the lunar city, where they could shop for artifacts and sample green cheese before boarding their ship back home. Some 400,000 people toured the Moon, including President William McKinley and most of his cabinet. When the exposition closed, Thompson took his attraction to Coney Island, calling it Luna Park.

The fascination with outer space probably explains the gullible reaction to the CBS radio broadcast of *The War of the Worlds* on October 30, 1938. Some of those who tuned in late made the mistake of concluding that the broadcast, by Orson Welles, was a news bulletin, not a story, and that angry Martians, at that precise moment, were incinerating seven thousand of America's finest soldiers. People rushed from their houses with wet clothes over their faces to protect them from a Martian gas attack. They fled in blind panic certain that the end was nigh. Police stations were inundated with frightened citizens seeking protection, and in Pittsburgh one woman had to be prevented from poisoning herself. The Welles broadcast was frightening because it overturned a comfortable American assumption about the superiority of Yankee technology. Most space tales, like those of Buck Rogers, had Americans traveling outward as conquerors, not being invaded, like victims.

When George Pal decided to make a Hollywood version of *War of the Worlds*, he set the film in present-day California. This was partly for reasons of economy (saving a bundle on props and costumes), but also to capitalize on the mania about flying saucers that had arisen since the Second World War. His film, *When Worlds Collide* (1951), was one of 133 science fiction movies Hollywood made between 1950 and 1957.[19] Whereas the previous generation of audiences had flocked to the cinema to watch white men massacre Indians, a new generation wanted battles with blobs, mutants, and things from another galaxy. One advantage of this new generation of villains was that it was not considered racist to do unspeakable things to an alien.

This obsession with outer space was part of the mania for all things scientific, and symptomatic of the assumption that science and technology could conquer all. It had many manifestations, from films, to television (*Tom Corbett: Space Cadet* debuted in 1950), to magazines like *Saturday Evening Post* and *Popular Science*, and naturally in comic books. The rocket offered an escape from the humdrum life of security and affluence, a chance for a new generation to imagine again the thrill of exploration and the challenge of colonisation. The rockets themselves might have been high tech, but the qualities required to pilot them were old fashioned. Buck Rogers succeeded not because of his equipment but because of true American grit. In this sense space promised a new frontier, another stab at the American Dream.

Since Americans thought themselves masters of technology, they concluded that the cosmos would be conquered as easily as the kitchen. At the 1939 World's Fair in Flushing Meadows, New York, large corporations paraded their latest high tech gadgets. The Ford exhibit was called "Road of Tomorrow," while General Motors dubbed its contribution "Futurama." Some of the items on display never made it into the shops or onto the roads, but most are now mundane household objects. The most popular exhibit was the "Democracity," which broadcast the message, none too subtly, that a free enterprise system would always remain at the cutting edge of technological change. In the modern metropolis, flat, straight superhighways would take citizens from the center to the suburbs, where modern houses would be built, or, more likely, assembled. These new homesteads would have air conditioning, stereophonic sound, central heating, two car garages, and family rooms where Mom, Dad, and the kids could entertain themselves in front of a box called television.

In the Transportation Pavilion an exhibit forecast a future of space exploration. From rocket ports, guests could board simulated flights to Mars and Venus, where they encountered primordial jungles full of strange beasts. It was commonly assumed that space, like the United States, would be democratized, that everyone would someday be able to board a rocket to the Moon or Mars, and that those celestial bodies would be granted statehood. Again, the rationale for the voyage was readily supplied by imaginative fantasies of what would be found.

The Second World War interrupted this vision, but also seemed to make it more likely. The part that technology played in the victory reinforced the idea that America would inevitably lead the way in the exploitation of new scientific frontiers. During the war, science brought fast new airplanes (even jets), rockets, missiles, and that most awe-inspiring invention, the atom bomb. Atomic power suggested that there would be endless energy to run high-tech household appliances. Lewis Strauss, postwar head of the Atomic Energy Commission, predicted that nuclear energy would eventually become so cheap that producers would not bother to charge for it. Abundant energy would, the propaganda claimed, "warm the cold, . . . feed the hungry . . . [and] alleviate the misery of the world." Scientists promised an atomic locomotive, nuclear greenhouses, and endless medical advances.[20] Engineers turned their minds to the possibility of using atomic explosions on large-scale construction projects, like building a second Panama Canal.

Science and progress were thought to be two sides of the same coin. Some people were anxious about what emerged from the laboratory, but most celebrated the new age of technology. The future promised revolutions in communication, medicine, engineering. Atomic physicists boasted of nuclear cars, submarines, and rockets. People would be made healthier, cities safer, and houses cleaner by the fruits of science. American entrepreneurial talent would take new technology into the homes of ordinary people, rendering everyone's life better.

Rockets seemed a logical expression of this new age. Where they might go, or what they might be used for, was immaterial. All that mattered was that they were fast and powerful. But, like the atom bomb, research on the rocket was hugely expensive and its financial potential not immediately apparent. Projects of this sort could not be carried out by earnest academics working diligently in small university laboratories. Nor could they be financed by private corporations who demanded a quick return on their investment. Only the government could build an atom bomb, and only the government could fly to the Moon.

This had enormous implications for the relationship between government and the private sector and, ultimately, for the liberal ethos. The need to compete with the enemy on the other side of the Iron Curtain implied a race for technological superiority which inevitably eroded the foundations of liberal democracy. On both sides, the Cold War was presented as a clash of diametrically opposed systems. Yet, as will be seen, the process of competition rendered the antagonists increasingly alike.

Two massive technocratic states emerged, competing with each other over the production of progress. Scientists, for whom academic freedom had once been a credo, now found themselves employees of the state, or, specifically, the military. The ivory tower gave way to the high-security establishment, while academics got used to a world of identity cards, security checks, tall fences, passwords, locked gates, and whispered secrets.

Dreamers and pragmatists met on the rocket pad. For the dreamers—men like Tsiolkovsky, Ehricke, and Bernal—space was a destination, a place to create or discover a perfect world. For the pragmatists—cold warriors like John Kennedy, Nikita Khrushchev, and Richard Nixon—space was more important for what it symbolized back on Earth than for what could be found in the sky. The dreamers supplied the inspiration, the pragmatists the cold hard cash. Each group needed the other. But while their goals intertwined, they never merged. The defining feature of the space age was that no one could ever agree on where the rockets should go.

2

Slaves to a Dream

For some people, rockets are erotic. The tall, slender, phallic tube sits on its pad while men who yearn for youth trade in techno-babble. The adventure appeals to most boys, some men, very few girls, and almost no women. Freud probably had a lot to say about this sort of thing, and would have said even more had he lived long enough to witness a thrusting V-2 raping the atmosphere. Most boys grow out of rockets around the time they become interested in girls. A small percentage don't, however, and they often become rocket scientists.

Just after the First World War, a group of these rocket-mad men in Germany got together with the aim of converting their adolescent dreams into physical reality. Their leading light was Hermann Oberth, originally from Transylvania. On May 3, 1922, while a student at Heidelberg University, Oberth wrote Robert Goddard requesting a copy of his controversial paper—the one pilloried in the *New York Times.* Goddard was flattered that someone was taking him seriously, but felt uncomfortable that his admirer resided in Germany. He was concerned about the German "tendency to turn inventions into weapons of war."[1]

A year later, Oberth published *The Rocket into Planetary Space.* Rather like its author, the book was a mixture of science and fantasy. Oberth had, at an early age, read Jules Verne, H. G. Wells, and virtually every other science fiction writer. His book proposed that liquid propellant rockets would allow man to escape the Earth's atmosphere. So far, so good. Then came the fantasy: rockets would lead inevitably to voyages to the outer planets and, eventually, permanent space stations.

The science was derivative, rather too much so for Goddard, who felt plagiarized. He had every reason to feel miffed, but he failed to understand a fundamental principle of space travel, namely that the thrust of a rocket is derived from equal parts fuel and publicity. Since Goddard shunned the latter, his paranoia limited the range of his dreams. The ability to promote oneself and encourage others to think big was essential in order to turn madcap ideas into real rockets.

On the strength of his book and his involvement with the film-maker Fritz Lang, Oberth drew a crowd of adoring rocket scientists. One of his admirers, the Austrian Max Valier, perceived a dramatic destiny: "The moment is here, the hour has come, in which we may undertake the attack on the stars with real prospects of results. It is clear that the armor of the Earth's gravity will not be lightly pierced, and it is to be expected that it will cost, to break through it, much sacrifice of time, money, and perhaps also of human life.[2]

Among the enthusiasts was an eighteen-year-old prodigy named Wernher von Braun. Like Oberth, he caught the space bug from reading science fiction novels as a young boy. These "filled me with a romantic urge. Interplanetary travel! Here was a task worth dedicating one's life to! Not just to stare through a telescope at the Moon and planets but to soar through the heavens and actually explore the mysterious universe! I knew how Columbus had felt."[3] What distinguished von Braun was that he was both a vivid dreamer and a brilliant engineer. He also possessed an uncanny talent for publicity, thus ensuring that people would shower him with money to pursue his fantasies.

Oberth's group of admiring rocket enthusiasts called themselves the Verein für Raumschiffahrt (Society for Space Travel, VfR), a name which indicates the nature and depth of their ambition. Members thought not only of shooting rockets into the sky, but also of attaching them to cars, boats, trains, sleds, and anything else that might be made to move fast. On May 23, 1928, a rocket car called the Opel-Rak 2 reached the death-defying speed of 125 miles an hour. The automobile magnate Fritz von Opel piloted it. Like Virgin Atlantic's Richard Branson, he understood that stunts made good publicity.

Before long, the German military grew interested in the powerful playthings of the VfR. Rockets were one category of weapon that had not specifically been mentioned when the Versailles diktats imposed strict limits on German armaments production, largely because no one took rockets seriously as a potential weapon. The Germans, however, recognized a loophole big enough to blast a missile through.

Professor-Colonel Karl Emil Becker, a widely respected ballistics expert, believed that rockets answered all the limitations as to range and destructive power posed by artillery, and, unlike airplanes, offered the possibility of carrying explosives to the enemy without the risk of losing a pilot en route. Another enthusiast was artillery captain Walter Dornberger who, in 1932, was assigned to investigate the feasibility of developing rockets for the Wehrmacht. He immediately knocked on the doors of the VfR. As he later explained, he found the young enthusiasts a bit too esoteric:

> The value of the sixth decimal place in the calculation of the trajectory to Venus interested us as little as the problem of heating and air re-generation in the pressurized cabin of a Mars ship. We wanted to ad-vance the practice of rocket building with scientific thoroughness. We wanted thrust-time curves of the performance of rocket motors. We wanted to establish the fundamentals, create the necessary tools, and study basic conditions.[4]

In other words, he wanted to be able to hit London with a missile. De-spite his more practical objectives, Dornberger fell instantly in love with the silver-tongued von Braun. On the strength of that attraction alone, the VfR was given a small grant to develop a prototype rocket.

Von Braun's first test shot failed spectacularly, but Dornberger was still sufficiently impressed to give the young aristocrat a position at the head of his new rocket team headquartered at Kummersdorf, about 100 kilometers south of Berlin. Von Braun packed his bags, telling his VfR colleagues that he'd been drafted. Leaving them behind in order to work for the god of war did not trouble him greatly, as he later ex-plained:

> In 1932 . . . when the die was cast, the Nazis were not yet in power, and to all of us Hitler was just another mountebank on the political stage. Our feelings toward the Army resembled those of the early aviation pioneers, who, in most countries, tried to milk the military purse for their own ends and who felt little moral scruples as to the possible fu-ture use of their brainchild. The issue in these discussions was merely how the golden cow could be milked most successfully.[5]

"We were interested in only one thing—the exploration of space," he added. Truth was always a fluid thing with von Braun; message and meaning carefully crafted to suit an audience. The refrain that he was only ever interested in space travel would prove handy throughout his life. It was probably true, but he was always pragmatic enough to real-ize that the quickest way into space would be aboard a military rocket. Arthur Rudolph, a brilliant rocket engineer but a rather vile Nazi, echoed these sentiments: "We didn't want to build weapons; we wanted to go into space. Building weapons was a stepping stone. What else was there to do but join the War Department? Elsewhere there was no money."[6]

Von Braun's acute sense of expediency does not adequately explain his unusual desire to join Nazi-affiliated organizations. He didn't seem to mind when the Deutscher Luftsport-Verbund flying club was taken over in 1933 by the N. S. Fliegerkorps (National Socialist aviation club). He also joined an SS horse-riding club, a Nazi-affiliated trade union, a hunting club, a welfare organization, and a National Socialist air raid protection corps.

The first rocket developed at Kummersdorf, the A-1, exploded magnificently. "It took us exactly one-half year to build," von Braun remarked, "and one-half second to blow up."[7] His team returned to their drawing boards and produced the A-2, first tested in December 1932. Two of these rockets, nicknamed Max and Moritz after German comic book characters, rose to 6,500 feet. Von Braun's reputation rose even further, as did his ego.

Enter Hitler. He and his colleague Hermann Goering saw enormous potential for rockets as machines of death. Von Braun wisely stopped talking about space travel, and tuned his radio to the Führer's frequency. By 1934, eighty engineers were working at Kummersdorf, a facility now identified with the suave, handsome, and ridiculously young von Braun, who had only just reached the ripe age of twenty-one. The apprentice had upstaged the sorcerer: Oberth was cast aside, mainly because his cosmopolitan virtues displeased the Gestapo. They did not like how he shared his ideas with rocketeers in other countries. The fact that he was not German also raised suspicions.

Before long, von Braun's ambitions, not to mention the range of his rockets, outgrew Kummersdorf. He needed a secluded place where a stray rocket would not end up in someone's sitting room. A new complex, called Peenemünde, was built on an isolated stretch of the Baltic coast at Usedom. Von Braun's staff moved there in the spring of 1937. The facility was every rocket scientist's fantasy: a purpose-built complex with virtually unlimited funding and a clear remit to produce rockets of ever increasing thrust. The only thing that might have annoyed the purist was the facility's purpose: there was never any question but that explosives would eventually be packed into the nose cones. Not long after decamping to Peenemünde, von Braun decided to join the

Nazi Party, a decision he claimed was inevitable. "I was officially demanded to join the National Socialist Party," he later explained. Since his rockets were attracting a great deal of official attention, he feared that "my refusal to join the party would have meant that I would have had to abandon the work of my life. Therefore, I decided to join. My membership in the party did not involve any political activity."[8]

The newest rocket, the A-3, failed in its first four tests, much to the dismay of Hitler, who had war on his mind. Von Braun's favorite gadget was the A-5, which in October 1938 reached a height of 8 miles. That fueled his dreams of exploring the heavens, but did not impress Hitler, who was more interested in range than altitude. With political pressure increasing, the team came up with the A-4, later known to the world as the V-2.* The single-stage rocket, fueled with alcohol and liquid oxygen, was about 46 feet tall and packed 56,000 pounds of thrust.

Progress at Peenemünde soon attracted the attention of Heinrich Himmler, the SS leader who was busily building his own empire, separate from the Wehrmacht and Luftwaffe. On May 1, 1940, von Braun was visited by a local SS colonel who happily informed him that Himmler wanted him to join. "I told him that I was so busy with my rocket work that I had no time to spare for any political activity," von Braun later explained. "He then told me, that my being in the SS would cost me no time at all. I would be awarded the rank of a 'Untersturmführer' [lieutenant] and it was a very definite desire of Himmler that I attend his invitation and join."[9] Worried that membership of the SS would obstruct his work for the military, von Braun consulted his mentor Dornberger. The latter advised him that it was not a good idea to annoy Himmler. He therefore joined, was promoted every year thereafter, but, he claims, never became remotely involved with the doings of the SS. He nevertheless enjoyed wearing his smart SS uniform whenever anyone important visited Peenemünde.

The first two tests of the V-2, in March and August 1942, were glorious failures. Though the rocket did not perform as intended, it still offered a tantalizing hint of its enormous potential. Then, on October 3 came a defining moment in the history of rocketry. "I kept my eyes glued to the binoculars," Dornberger recalled; "it was an unforgettable sight. In the full glare of the sunlight the rocket rose higher and higher. The flame darting from the stern was almost as long as the rocket it-

* The name comes from Vergeltungswaffen, or "revenge weapon."

self. . . . The air was filled with a sound like rolling thunder."[10] The V-2 eventually reached an altitude of 60 miles, taking it to the edge of space. Ehricke, a member of the elite team, was moved to philosophize:

> It looked like a fiery sword going into the sky. There came this enormous roar and the whole sky seemed to vibrate; this kind of unearthly roaring was something human ears had never heard. It is very hard to describe what you feel when you stand on the threshold of a whole new era; of a whole new age. . . . It's like those people must have felt— Columbus or Magellan—that for the first time saw entire new worlds and knew the world would never be the same after this. . . . We knew the space age had begun.[11]

Dornberger showered congratulations like confetti. "We have invaded space with our rocket," he announced triumphantly, "and for the first time have used space as a bridge between two points on Earth; we have proved rocket propulsion practicable for space travel. To land, to sea and [to] air may now be added infinite empty space as an area of future intercontinental traffic." But then came the punch line: "So long as the war lasts, our most urgent task can only be the rapid perfecting of the rocket as a weapon."[12]

On July 7, 1943, von Braun and Dornberger met Hitler to talk about the V-2. They told him that the rocket could carry 1,000 kilos of explosives to London and would be impossible to intercept. After being shown a film of the V-2 in operation, Hitler replied that he wanted 2,000 rockets a month. In order to meet such a demand, von Braun needed a huge factory and priority over scarce raw materials and fuel. Since Hitler had already decided that the V-2 would win the war, the request posed no problem. Albert Speer, Minister of Supply, was instructed to give the rocket program top priority. "This will be retribution against England," Hitler said. "We will force England to her knees. The use of this new weapon will make any enemy invasion impossible."[13]

Hitler was in fact thinking bigger. Why, he asked, couldn't the team make a rocket capable of carrying a ten-ton payload? Dornberger explained that such a rocket would take years to develop. Hitler didn't like that answer. "What I want is annihilation!" he shouted. Dornberger replied: "We were not thinking of an all-annihilating effect." "You!" Hitler spat. "No, you didn't think of it, I know. But I did!"[14]

The Führer had fallen victim to rocket fever. Let's suppose that 2,000 rockets could indeed have been produced each month and that each of those rockets could have hit targets in England. (Both are very reckless assumptions.) That would have meant roughly 2 million kilos of explosive delivered to Britain every month. By the time the V-2 was deployed in 1944, Allied bombers were dropping around 3 million kilos on Germany *every day*, yet were not destroying German morale. As it turned out, just over 1,300 rockets were fired at London by the end of the war, causing 2,700 deaths and over 6,000 serious injuries. Another 1,265 rockets fell on Antwerp. Tragic as those losses were, they did not remotely affect Allied morale and in fact probably stiffened resolve. One estimate places the total cost of the project at RM 2 billion, equivalent at the time to $500 million. That's one-quarter of what the United States spent on the Manhattan Project.[15] "Our most expensive project was also our most foolish one," Speer later admitted. "Those rockets, which were our pride and for a time my favorite armaments project, proved to be nothing but a mistaken investment."[16]

On March 15, 1944, von Braun was arrested by the Gestapo on charges that he was wasting Reich resources on his dream of space travel. While he was indeed interested in exploring the cosmos, the charges on this occasion lacked foundation. Thanks to the intervention of Speer, he was released after two weeks. While the experience was undoubtedly frightening, it proved useful after the war when von Braun needed evidence to suggest that he was never an ardent Nazi, nor an enthusiastic missile maker.

Von Braun always assumed a benign innocence when it came to discussing his Faustian deal with Hitler. "It seems that this is another demonstration of the sad fact that so often important new developments get nowhere until they are first applied as weapons," he told his colleagues. One of those colleagues had a different take: "Don't kid yourself—although von Braun may have had space dust in his eyes since childhood—most of us were pretty sore about the heavy allied bombing of Germany—the loss of German civilians, mothers, fathers, or relatives. When the first V-2 hit London, we had champagne. Why not? We were at war, and although we weren't Nazis, we still had a Fatherland to fight for."[17]

V-2 production, as Speer suggested, placed intolerable demands on Germany's already overstretched labor supply. The problem was solved by taking slaves from concentration camps. Von Braun's team were already inclined to tap this source even before Hitler made his extraordinary demand. Rudolph had observed the use of slaves at the Heinkel aircraft works and felt that "the employment of detainees in general has had considerable advantages over the earlier employment of foreigners, especially because all non-work-related tasks are taken over by the SS and the detainees offer greater protection for security."[18] He recommended their immediate employment on the rocket program, formally requesting 1,400 slave laborers from the SS on June 2, 1943. This request could not have been made without von Braun's full knowledge. Two months later, Hitler ordered Speer to use concentration camp labor to build the new rocket factories, and eventually to build the rockets themselves.

An underground facility at Mittelbau Dora, near Nordhausen in eastern Germany, was built to manufacture the rocket components. Construction and organization were supervised by SS General Hans Kammler, one of Heinrich Himmler's favorites and a truly vile man. Dora was essentially a satellite camp of Buchenwald, from which able-bodied laborers were drawn and then worked to death. The first group of 107 prisoners arrived in August 1943. They set to work carving out new tunnels to enlarge an existing storage depot. Over the following six months, an additional 12,000 prisoners arrived to work in the damp caverns. Conditions were atrocious, survivors rare. There was no heat, no ventilation, no sewerage, no sinks or tubs, no running water. Thirsty prisoners were often forced to drink the putrid water oozing from the rock walls of the tunnel. Malnutrition and disease were rife. "The sanitation was totally inhuman," commented Angela Fiedermann, a former inmate. "The temperature was eight or nine degrees Celsius . . . and humidity was 90 percent. They died like flies."[19] The work itself was exhausting, backbreaking, and dangerous, with a constant threat of explosions, cave-ins, and gas poisoning. Prisoners worked upwards of seventy-two hours per week, existing on a mere 1,100 calories per day. A crematorium had to be built in order to handle the steady stream of prisoners who died or were murdered on the job. Probably 20,000 perished building the V-2s, and countless more in plants where fuel for the rockets was processed.

Dr. A. Poschmann, medical supervisor to the Armaments Ministry, thought the tunnels a reasonable facsimile of Dante's Inferno. On December 10, 1943, on his urging, Speer visited the facility and noted:

Expressionless faces, dull eyes, in which not even hatred was discernible, exhausted bodies in dirty gray-blue trousers. At the approach of our group, they stood at attention upon hearing a cutting command and held their pale blue caps in their hands. They seemed incapable of any reaction. . . . The prisoners were undernourished and overtired; the air in the cave was cool, damp and stale and stank of excrement. The lack of oxygen made me dizzy; I felt numb.

Speer immediately ordered Kammler to construct a more hygienic above-ground camp for prisoners and to improve the food and sanitary facilities. Or so he claims. If these orders were indeed given, they were not carried out. A short time later, Speer formally commended Kammler for his work. "In an almost impossible short period of two months," he remarked, the SS officer had transformed "the underground facilities from a raw state into a factory." The achievement outshone every "remotely similar example anywhere in Europe and is unsurpassable even by American standards."[20]

Those who survived the terrible conditions had still to cope with the capricious brutality of the guards. Prisoners were executed for no other reason than to maintain an atmosphere of terror. Sadistic guards tortured detainees simply for fun. Random reprisal killings inevitably followed suspected acts of sabotage. For the workers, sabotage (which sometimes took the simple form of urinating on delicate wiring) made sense since death seemed inevitable. This helps to explain why the V-2s had such a low success rate.

When sabotage was suspected, retribution was brutally swift. René Steenbeke, a captured Belgian army officer, recalled seeing "51 prisoners . . . hanged, their hands behind their backs, a piece of wood in their mouths, hanged in groups of about 12. They could see their comrades being killed before them and they had to watch."[21] The huge cranes were used as makeshift scaffolds, even though this slowed production. Hans Friedrich, a civilian manager at the factory, recalls phoning Rudolph to ask how long the bodies should be left dangling from the crane. The latter replied that they should be left to hang for the last six hours of one shift, and the first six hours of the next, in order that the entire crew could benefit from the lesson.

In early 1945, the number of prisoners peaked at 40,000. At that time, Dora was producing 690 rockets and killing 2,000 prisoners per month, some of them executed, but most of them simply worked to death. On April 11, the facility was captured by a detachment of the American 3rd Armored Division. In one of his final acts, Hitler had ordered Himmler to execute all remaining prisoners, but this proved impossible. Nevertheless, the Americans found "hundreds of corpses . . . sprawled over the acres of the big compound. More hundreds filled the great barracks. They lay in contorted heaps, half stripped, mouths gaping in the dirt and straw, or they were piled naked like cordwood."[22] According to one count, over five thousand corpses were found.

Steenbeke feels that the facility should be preserved as a reminder of where the space age began. "Everything that is now in space had its origins here, not in America or Russia," he says. "This is where a new science started, but it is also where science and death met."[23] Jean Michel, another rare survivor, remarked: "I could not watch the Apollo mission without remembering that that triumphant walk was made possible by our initiation to inconceivable horror."[24]

And what of von Braun? He was fully aware of the extent to which his project relied upon slave labor. His younger brother, Magnus, worked in the factory on gyroscope production and kept him informed of conditions there. Von Braun also, on a number of occasions, went to Buchenwald to shop for slaves with specialist skills. On January 24, 1944, he toured Dora and presumably witnessed virtually the same conditions Speer had seen six weeks before. Admittedly, von Braun would have preferred not to use slaves—not because it was immoral, but because it was inefficient. He found them terribly undependable workers: they slowed production, encouraged sabotage, and produced an unreliable rocket.

In 1947, von Braun declared under oath that he visited Dora fifteen to twenty times during the war. He maintained that, if at first the facilities were primitive, over time significant improvements were made on his urging, including the installation of air conditioning. Interviewed again in the 1970s, he admitted that it was "a pretty hellish environment. The conditions there were absolutely horrible." On that occasion, he recalled "several" visits.[25] For reasons of ambition or self-protection, he never complained about the conditions during the war. After it, the issue was immaterial, except to the extent that it might tarnish his carefully crafted reputation. Dornberger showed even less concern. "I had made rockets my life's work," he remarked. "Now we had to prove that their time was come, and to this duty all personal considerations had to be subordinated. They were of no importance."[26]

Von Braun had the instincts of a rat: he knew precisely when to abandon a sinking ship. In January 1945, when the Russian Army was closing on Peenemünde, he gathered together his senior staff. He explained that he had no intention of fighting to the death for Nazism. Instead, it seemed best to surrender to that power which could be expected to treat them best. As one member of his team remarked: "We despise the French; we are mortally afraid of the Soviets; we do not believe the British can afford us; so that leaves the Americans."[27]

Over the next few months, Von Braun devised an intricate plan for exporting the entire apparatus of rocket manufacture to the United States. Blueprints, machinery, engineers, and scientists were packed into trains, trucks, and cars and moved westward, the destination Oberammergau. The goal was to get as far as possible away from the advancing Russians, but also to distance themselves from Dora, which might otherwise raise awkward questions. Important papers were carefully hidden in a mineshaft, protected by booby trap explosives. "These documents were of inestimable value," one of von Braun's colleagues later explained. "Whoever inherited them would be able to start in rocketry at that point at which we had left off. . . . They represented years of intensive effort in a brand-new technology, one which, all of us were still convinced, would play a profound role in the future course of human events."[28]

On May 2, 1945, Magnus von Braun was out on his bicycle when he ran into an American patrol. He told the startled Yanks that an entire team of German rocketeers were holed up a few miles back. Arrangements were quickly made for von Braun and his colleagues to cross the American lines later that evening. Von Braun quickly made it clear that he was fully prepared to cooperate and would show his captors precisely where the remaining rockets could be found. The Americans ended up with the senior engineers, the plans, a great deal of the hardware, and most of the remaining weapons. The booty was packed into boxcars and shuttled westward. The first train, with forty freight cars, left Nordhausen on May 22. For the next nine days, a train of equal size, packed with rocket parts, embarked for Antwerp. The last train left Nordhausen a few hours before the Soviets were due to take over the area. Meanwhile, the cave filled with documents, located in the British sector, was emptied just minutes before the Allies arrived. Fourteen tons of paperwork was extracted, including over 3,500 detailed reports and over 500,000 blueprints.

The Americans found von Braun a jovial chap eager to tell them all about his rockets. He basked in the Americans' attention, and only occasionally complained about being held in a camp surrounded by barbed wire. The fact that he had shot missiles at England and built them with slave labor was by this time a small detail, a minor matter confined to the past. He nevertheless boasted that, given two more years, he could have single-handedly won the war, especially since he had been working on a rocket with a range sufficient to hit New York. A paper detailing his work was quickly prepared by von Braun and given to the Americans as a teaser of what he could do. Working on weapons was, he claimed, "an intermediate solution conditioned by this war." Now that the ugly business of fighting was behind them, "it will be possible to go to other planets, first of all the Moon."[29]

A little over one hundred of the thousand technicians and engineers who made the trek with von Braun were eventually taken to the United States, not as prisoners but as honored guests. As one of von Braun's colleagues recalled: "I realized more and more that [the Americans] wanted something from us. And of course that has to be paid for. We had to sell ourselves as expensively as possible."[30] Von Braun played a clever game. He knew that the Americans were desperate. He responded with grace and bonhomie, but behind the avuncular exterior lurked the mind of a drug pusher. He gave the Americans free samples of rocket know-how, realizing full well that once they developed an addiction, he could demand whatever price he wanted.

The exodus of the rocket scientists was part of a larger program, called Operation Overcast, for the wholesale transfer of German expertise to the United States. The War Department tried to cast the policy in the best possible light, in case some people at home had scruples about employing ex-Nazis. On October 1, 1945, the department issued the following press release:

> The Secretary of War has approved a project whereby certain understanding German scientists and technicians are being brought to this country to ensure that we take full advantage of those significant developments which are deemed vital to our national security. . . . Throughout their temporary stay in the United States these German

scientists and technical experts will be under the supervision of the War Department but will be utilized for appropriate military projects of the Army and Navy.[31]

The aim was threefold: to prevent German rearmament, to keep the Soviets from making use of valuable expertise, and, finally, to use German know-how for the benefit of the United States. As the War Department implied, the original intention was that the scientists would be brought to the United States, be extensively de-briefed, and then be returned to Germany when the situation seemed sufficiently stable. Von Braun, however, had a different idea. "Our primary concern was 'stability' and continuity," he later explained. "We were interested in continuing our work, not just being squeezed like a lemon and then discarded."[32]

The policy caused a few ripples back home. Albert Einstein, in conjunction with the religious leader Norman Vincent Peale and union organizer Philip Murray, protested to President Truman. "We hold these individuals to be potentially dangerous carriers of racial and religious hatred," they argued. "Their former eminence as Nazi party members and supporters raises the issue of their fitness to . . . hold key positions in American industrial, scientific, and educational institutions." Rabbi Stephen Wise pointed out how ironic it was that former victims of concentration camps were denied admission to the United States, while their former captors were welcomed with open arms. "I have never thought we were so poor mentally in this country that we have to import those Nazi killers to help us prepare for the defense of our country," complained John Dingell, a Democratic congressman from Michigan.[33] A few moral scruples would not, however, get in the way of national security.

Before long, American officials decided that the Germans were too valuable to allow them to return to their homeland. In 1946, a move was initiated to permit them to stay permanently in the United States. This caused a problem with U.S. immigration, which denied permanent residence to members of fascist organizations or to those who had actively opposed American war efforts. The War Department's wishes were nevertheless granted, and a new system, called Operation Paperclip,* was authorized by President Harry Truman on September 6, 1946.

---

* The name was said to have originated because immigration forms were paper-clipped to the papers collected in the investigation of potential recruits.

Somewhat disingenuously, a new order proclaimed that "no person found by the Commanding General, USFET (U.S. Forces European Theater) to have been a member of the Nazi Party and more than a nominal participant in its activities, or an active supporter of Nazism or militarism shall be brought to the U.S."[34]

Paperclip was administered by the Joint Intelligence Objectives Agency (JIOA), operating under the Joint Chiefs of Staff (JCS). Investigations into the background of the individuals chosen for "adoption" were conducted by the Office of the Military Government U.S. (OMGUS). Case files were then read by the State Department, which pronounced upon suitability. For the State Department, the foremost issue was whether the candidate was considered to have been an "ardent Nazi." Samuel Klaus, their representative on the JIOA, pressed for thorough investigations of every individual. He was particularly concerned that many specialists on the War Department's wish list had lied about their past and that discrepancies had not been properly investigated.

War Department ambitions and State Department scruples inevitably collided. At a meeting on May 26, 1947, Herbert Cummings, assistant chief of the State Department's Bureau of Foreign Activity Correlation, "hit the ceiling" when he found discrepancies between OMGUS investigations and reports submitted to the State Department. Someone was cooking the records. "We got into several rounds because it looked like they [the War Department] were trying to dump," Cummings later revealed.[35]

Navy captain Bosquet Wev, director of the JIOA, found the attitude at the State Department annoying and naïve. Fearing that any rejected scientists would end up in the hands of the Soviets, he complained that the State Department was "beating a dead Nazi horse" by demanding rigorous security checks.

> It is a known fact that any German who lived in Germany during the war and who possessed any capabilities whatsoever, was a member of some affiliation to the Nazi Party. Otherwise he was placed in a concentration camp. The determining factor lies in the question of just what constitutes an active Nazi. Furthermore, . . . the fact that an individual was a loyal German or was affiliated with the Nazi Party does not in itself imply that he is now a security threat to the United States.

Wev warned that to return the scientists to Germany, where they could be poached by the Russians, "presents a far greater security threat to this country than any former Nazi affiliations which they may have had or even any Nazi sympathies that they may still have."[36] In a stinging rebuke of the State Department written in March 1948, he argued: "If we are to face the situation confronting us with even an iota of realism, Nazism no longer should be a serious consideration from a viewpoint of national security when the far greater threat of Communism is now jeopardizing the entire world."[37]

Unable to reach a consensus with the State Department, the JIOA decided to sanitize unfavorable OMGUS reports. On November 18, 1947, Wev's assistant Walter Rozamus sent a memo to the Intelligence Division General Staff Army asking them to withhold dossiers containing incriminating evidence. Ten days later, an official memo formally asked army investigators "to re-evaluate these reports." Since the individuals in question "were not considered to be potential security threats to the United States . . . their classification as ardent Nazis should be revised since such classification is a bar to immigration."[38] Rozamus was particularly concerned about the case of von Braun who, according to OMGUS, was "a potential security threat to the United States and . . . will be wanted for denazification trial in view of his party membership."[39]

On December 4, 1947, Wev asked the European Command Director of Intelligence to review the files of fourteen individuals, among them von Braun, with a view to producing a more palatable report. After this second review, von Braun's report read:

> No derogatory information is available on the subject individual except NSDAP records, which indicate that he was a member of the party from 1 May 1937 and was also a Major in the SS, which appears to have been an honorary commission. The extent of his Party participation cannot be determined in this Theater. Like the majority of members, he may have been a mere opportunist. Subject has been in the United States more than two years and if, within this period, his conduct has been exemplary and he has committed no acts adverse to the interests of the United States, it is the opinion of the military governor, OMGUS, that he may not constitute a security threat to the United States.

Arthur Rudolph's file originally read: "100% NAZI, dangerous type, security threat. . . !! *Suggest internment.*" After Wev intervened, the cleaned-up report stated: "Nothing in his records indicating that he was a war criminal, an ardent Nazi or otherwise objectionable." Overwhelming evidence linking Rudolph to the mistreatment of Dora inmates, including the mass hangings, was simply filed away. Linda Hunt, a journalist who later investigated the activities of the JOIA, found that all of the 130 OMGUS reports she examined had been altered.[40]

Among those recommended for permanent residence was Siegfried Ruff, who had headed the Department for Aviation Medicine at the German Experimental Institute. Ruff's experiments on the effect of high-altitude flying are thought to have killed up to eighty Dachau inmates who were placed in low-pressure chambers in order to simulate conditions at 68,000 feet. After the war, the Americans paid Ruff to write reports derived from that research. Another person given a clean bill of health was Dr. Hubertus Strughold, onetime director of the Luftwaffe's aeromedical institute, who conducted experiments similar to those of Ruff. An April 1947 intelligence report clearly stated that Strughold's "successful career under Hitler would seem to indicate that he must be in full accord with Nazism." Yet on a 1952 civil service application form he denied any connection with the Nazi Party, and was not taken to task for this omission. His work in the United States was so widely respected that in 1977 a building at the Air Force School of Aerospace Medicine was named in his honor.[41]

Walter Schreiber worked for the American space program until 1952, when his past caught up with him. When evidence surfaced that he had participated in the Nazi extermination programs and medical experiments, the JIOA arranged a visa for him to go to Argentina, found him a job there, and even threw in a free flight. Rudolph, who worked on the Saturn V rocket, escaped exposure until October 1984, by which time the task of beating the Russians to the Moon had long been accomplished. In that year, the Justice Department's Office of Special Investigations suddenly uncovered "irrefutable" evidence of his "complicity in the abuse and persecution of concentration camp inmates who were employed by the thousands as slave laborers under his direct supervision." Rather than face charges, he surrendered his American citizenship and moved back to Germany. He claimed that he knew nothing of killings at Dora and "only deduced there must have been deaths among the prisoners because they were exposed to the same conditions as my Germans and I myself was."[42]

In 1947, von Braun informed his American hosts that he wanted to travel back to Germany in order to marry his eighteen-year-old cousin. They funded the trip on the understanding that he would immediately bring his wife back to the States. American secret service agents were guests at the wedding and, during the honeymoon, lurked quietly in the shadows, ever vigilant in case the Russians decided to pounce. Two years later, as quietly as possible, von Braun and his fellow Germans were loaded into a bus and taken across the U.S. border at El Paso. The bus then did a U-turn and headed back through the border control station, so that the passengers could all be given unlimited entrance visas, which they then used to apply for American citizenship.

3

What Are We Waiting For?

Stalin had calculated that his forces would reach Peenemünde before the Americans, but when they arrived the cupboard was bare. "There was hardly a German sufficiently competent to talk about the V-2 and other big stuff," Grigory Tokady, a Russian rocket scientist sent to investigate immediately after the German defeat, later revealed. "Many, almost all, claim[ed] to be V-2 experts . . . [but they] displayed the typical characteristics of a second-rater." Stalin could hardly contain his fury. "This is absolutely intolerable," he spat. "We defeated Nazi armies; we occupied Berlin and Peenemünde, but the Americans got the rocket engineers. What could be more revolting and more inexcusable? How and why was this allowed to happen?"[1]

The Russians knew a bit about rockets. In the 1930s, Sergei Korolev had been at the forefront of Soviet experimentation. His Group for the Investigation of Reactive Motion was essentially a Russian version of the VfR, testing liquid-fuel devices. At the same time, Ivan Kleimenov and his deputy, Georgiy Langemak, made significant advances at the Scientific Research Institute Number 3, a government-sponsored effort. Progress was, however, threatened when Kleimenov and Langemak fell victim to a Stalinist purge in 1937 and were executed on trumped-up charges of spying for the Germans. In the following year, the secret police came for Korolev, who was thrown in the Kolyma gulag in eastern Siberia on charges of sabotage. "Our country doesn't need your fireworks," his interrogator told him.[2]

War brought a more pragmatic attitude, with the result that in 1942 Korolev was allowed to conduct rocket research from within prison. Within two years his team had developed the RD-1, a liquid fuel rocket engine eventually incorporated into fighter and bomber production. Delighted with this success, Stalin pardoned Korolev and his team or, more precisely, announced that they had been "rehabilitated."

After familiarizing himself with the state of German rocketry at Peenemünde, Tokady concluded that the USSR was the equal of Germany in rocket theory, but significantly behind in practice. He surmised that, if energy could be properly directed, there was no reason why the Soviets would not be able to duplicate the success of the V-2. The main reason the Soviets were behind the Germans at the end of the war was not because they did not understand how to build long-range rockets, but rather because they had chosen not to do so. They had concentrated their industrial effort into more immediate, practical, and down-to-Earth weapons. In view of the German experience, that was an intelligent decision.

Korolev was commissioned a colonel in the Red Army and sent to Germany to pick over what was left of Wernher von Braun's rocket industry. He collected some 150 personnel and took them back to Russia, where they remained until around 1953. American journalists would later joke that the space race boiled down to "their Germans" against "our Germans." In fact, that was unfair, since Korolev never allowed his Germans to be incorporated into the program to quite the same extent that they were in America. Their brains were picked of all useful information, whereupon they were sent packing. Germans were not allowed to develop with the program or to offer creative direction. The Russian space program was a lot more Russian than the American was ever American.

Soviet necessity became the mother of invention. Since the Americans had been able to pilfer a large number of V-2 rockets, they made do with these well into the 1950s. The Russians, in contrast, quickly ran out of surplus V-2s and were forced to make their own, grafting German know-how onto home-grown research. The result was the Pobeda rocket, a vast improvement on the V-2, and thus a leap ahead for the Russians.

Scratch beneath the surface, and one quickly finds that the giants of Cold War rocket technology had more in common than in contrast. Von Braun and Korolev were both visionaries who dreamed of putting objects into orbit, and men into space. Both knew, however, that their fantasies could never be realized unless they could be sold as practical programs for the defense of the state. Both therefore willingly sold their souls to arms production, while telling themselves that a spacecraft is merely a missile by another name.

Stalin wanted missiles for the simple reason that he needed a way to deliver his nuclear weapons (once they were developed) to the United States. Since he did not have allies on the borders of his enemy (as the United States did) he could not rely on bombers to carry his weapons. He wanted instead to leapfrog the U.S. advantage in weaponry by countering it with an advantage in delivery systems. In April 1947 he gathered together rocket scientists and military leaders and ordered them to develop missiles capable of hitting the United States. "Do you realize the tremendous strategic importance of machines of this sort?" he asked. "They could be an effective strait jacket for that noisy shopkeeper Harry Truman. We must go ahead with it, comrades."[3] "The Soviet Union felt naked, unprotected, surrounded everywhere by American nuclear forces," Nikolai Detinov of the Soviet Defense Ministry later recalled. "When we . . . developed our own ballistic missiles, although we had very few, we realized that it . . . acted as a counterbalance."[4] Rockets were also cost effective—less expensive, easier to maintain, and less vulnerable than bombers. To the Soviets, the bomber seemed a decidedly old-fashioned way to deliver death.

In public, however, they tried desperately to maintain the moral high ground, at least until their rockets were ready. This meant attacking the United States for its "Hitlerite ideas and technicians." Writing in the *New Times,* an English-language magazine published in Moscow, a Soviet spokesman condemned the "sect" within the United States which pursues "the ruthless, ghoulish doctrine of the Hitler marauders, which in the end proved so fatal to themselves. . . . The Nazi bombardment of London with V-2 weapons proved how futile such attempts are. . . . The argument . . . that intercontinental weapons will be a 'grand strategic weapon' is, from the military point of view, a wild utopia. Actually, it is a piece of deliberate bluff, an attempt at extortion by intimidation."[5] Behind the rhetoric of outrage, Korolev was building the most powerful rocket the world had ever seen.

Meanwhile, the V-2s, repainted with American colors, were fired from test sites at White Sands, New Mexico. The first was launched on April 16, 1946, with another sixty-four fired over the following six years. They were not, strictly speaking, weapons tests, since the rockets lacked explosive warheads, but the data collected were incorporated into what became a wide-ranging project to produce ballistic missiles. At the same time, however, the tests doubled as scientific experimentation, with cameras, Geiger counters, and other scientific paraphernalia occasionally placed in the nose cones. In 1946, for instance, a sensor in a V-2 demonstrated the existence of the ozone layer. Sometimes, however, even the science was militaristic. While nose-cone cameras provided nice snapshots of Earth, they also underlined the possibility of using satellites for reconnaissance.

For von Braun, life was comfortable, but not fulfilling. While his dreams were limitless, those of his new masters were decidedly finite. The Truman administration, keen to make economies, cut the military budget from $81.5 billion to $13.1 billion in the two years after the war. Over the same period, the amount allocated to the Army Air Corps for rocket research fell from $29 million to $13 million. Since the White House felt confident that the atom bomb provided a very cheap way to demonstrate American military power, it seemed safe to cut other programs. In this climate of frugality, purely speculative projects had little hope of thriving. Rocket research, in other words, would attract money only if it could produce cheap and practical devices capable of improving the nation's defenses. The benefit of the V-2 was immediately obvious—a beefed-up version would make a useful intercontinental ballistic missile. Exploring space, however, seemed to have no practical benefit, at least among the bean counters at the War Department.

In order for von Braun to be given the green light to go into space, he (or someone) needed to demonstrate that space offered unique potential for national security. Opinion on this subject was deeply divided. In 1945, a team commissioned by the navy promoted the idea of an Earth-circling satellite capable of relaying valuable information back to Earth. The group had obviously been influenced by captured German documents containing proposals to this effect. But this enthusiasm was not mirrored by the higher brass who could not immediately see how a satellite would make the job of fighting a war at sea any easier. The report was filed.

The army was slightly more enthusiastic. Army aviators, who would shortly split off to form the air force, asked their civilian suppliers to submit proposals for a satellite. The most attractive response came from Douglas Aircraft, which, it so happened, had only recently set up a research and development group in Santa Monica, California, called RAND. Since that group's main function was to think about all aspects of nuclear warfare, the satellite fit in rather nicely with its prognostications.

In a report completed in May 1946, RAND predicted that a very useful device could be developed and put into orbit for a mere $150 million. What it would actually do was not entirely clear. A nod was given in the direction of science: "A satellite vehicle with appropriate instrumentation can be expected to be one of the most potent scientific tools of the Twentieth Century," it was argued. More important were the political implications. "The achievement of a satellite craft by the United States would inflame the imagination of mankind, and would probably produce repercussions in the world comparable to the explosion of the atomic bomb."[6] In line with that optimism, the group predicted that the task could be achieved as early as 1951.

Louis Ridenour, author of that first report, was stubbornly practical, as befits a RAND analyst. But, toward the end, even he could not resist flights of fancy typical of those who look to the sky:

> The most fascinating aspect of successfully launching a satellite would be the pulse quickening stimulation it would give to considerations of interplanetary travel. Whose imagination is not fired by the possibility of voyaging out beyond the limits of our Earth, travelling to the Moon, to Venus and Mars? Such thoughts when put on paper now seem like idle fancy. But, a man-made satellite, circling our globe beyond the limit of the atmosphere is the first step.

Subsequent steps, Ridenour thought, would follow in quick and logical progression, like an escalator to the Moon. "Who would be so bold as to say that this might not come within our time?"[7]

What is clear from RAND's musings is that the space race, which hadn't officially begun, was already being driven by propagandistic concerns. Practical applications for satellites were secondary to their value as symbols of cultural virility. A subsequent report, by J. E. Lipp, completed in February 1947, drove this point home:

> Since mastery of the elements is a reliable index of material progress, the nation which first makes significant achievements in space travel will be acknowledged as the world leader in both military and scientific techniques. To visualize the impact on the world, one can imagine

the consternation and admiration that would be felt here if the U.S. were to discover that some other nation had already put up a successful satellite.

The author went on to argue that a race in space was inevitable and that the country (namely the Soviet Union) that was "behind" on Earth would see space as a way to catch up:

> Since the United States is far ahead of any other country in both airplanes and sea power, and since others are abreast of the United States in rocket applications, we can expect strong competition in the latter field as being the quickest shortcut for challenging this country's position. No promising avenues of progress in rockets can be neglected by the United States without great danger of falling behind in the world race for armaments.[8]

For the moment, Lipp's warning went unheeded. The fact that they had so cleverly snatched von Braun and the majority of his team bred a dangerous complacency among Americans. They thought that they had a winning hand or, put differently, that the space race had been won even before it began.

Complacency was encouraged by deeply imbedded cultural stereotypes. Americans assumed that the Soviets did not have the technical brains to devise rockets, nor the engineering prowess to build them. Vannevar Bush, one of the leading lights in the effort to build the atom bomb, and a man to whom Truman listened intently, argued in 1949: "The type of pyramidal totalitarian regime that the Communists have centered in Moscow . . . is not adapted for effective performance in pioneering fields, either in basic science or in involved and novel applications. . . . Hence it is likely to produce great mistakes and great abortions." This also meant that "no other nation will have the atomic bomb tomorrow."* To complete his hat trick of daft predictions, Bush maintained that the ballistic missile "would never stand the test of cost analysis. If we employed it in quantity, we would be economically exhausted long before the enemy."[9]

The RAND reports made thinking about satellites respectable, but did not lead to a flood of support for practical ventures. In October 1947,

---

* True, not tomorrow, but within a few months of that prediction.

the Committee on Guided Missiles of the Joint Research and Development Board made favorable noises, but recommended that an actual program of satellite development be delayed until a practical military use could be defined. A navy report in the following year concluded that, while the technology to launch a satellite was conceivable, no "military or scientific utility commensurate with the expected cost" had been proven.[10] Clearly, those who controlled the purse strings were not yet prepared to spend millions on a propaganda symbol in the sky. A thrust forward in satellite development would have required enthusiasm from *both* the army and the navy. That was sorely lacking. Both were obsessed with protecting their own turf at a time when competition for funding was intense.

The National Security Act of August 1947 had created the Department of Defense, replacing the old War Department. The secretary of defense, James Forrestal, filed his *First Report* on December 29, 1948, and revealed that all three branches of service (the air force had been established the previous year) had bodies studying the feasibility of satellites. This sparked an angry reaction among ordinary Americans, who objected to what seemed blatant boondoggling. The chastened government reacted by axing the High Altitude Test Vehicle (HATV) program, conducted under the auspices of the navy. The HATV, which had a projected launch date in 1951, was designed by the Glenn L. Martin Company, and, in retrospect, seems to have been significantly ahead of anything the Soviets were doing at the time.

American complacency was shattered in September 1949 when a weather plane accidentally picked up indications of an atomic blast in the Soviet Union. The fact that the Russians had detonated an atomic bomb years, if not decades, ahead of "expert" predictions changed the climate of the Cold War overnight, undermining American confidence about the safety of their nuclear umbrella. They consequently felt a need to move into new areas of defense where their presumed technological superiority would give them an advantage. This provided an obvious boon for those working in rocketry, because satellites seemed to offer the possibility of keeping tabs on what the Soviets were doing.

Once both powers possessed nuclear weaponry, methods of delivery became much more important. Americans understood that Stalin was keen on developing intercontinental ballistic missiles, and felt obliged to follow the same path. For von Braun, this was both a blessing and a curse. Greater emphasis was placed on rocket development, but the overriding purpose was missiles, not space. Von Braun had come full circle. He was once again doing what he had done at Peenemünde, making rockets designed to deliver death to an enemy city. His new weapon, the Redstone rocket, was the muscular son of the V-2—faster, more dependable, more accurate, and (because of its nuclear warhead) a great deal more powerful. Somewhat ironically, those rockets would eventually be deployed in Germany, but this time aimed eastward, toward the Soviet Union.

In early 1950, von Braun and his team were moved to the Redstone Arsenal in Huntsville, Alabama, with a specific remit to develop ground-to-ground ballistic missiles equipped with nuclear warheads. Inevitably, the place became known as "Hunsville." For the Germans, the New Mexico desert around Fort Bliss had seemed like punishment. Their transfer to Huntsville came like a reward for five years of loyal service. Many felt that their penance had been served and they could now start to rebuild their lives in an environment more suited to their tastes. Very few expressed a desire to return to Germany, and even fewer acted on that desire. They had decided to become Americans, to buy homes, play baseball, make pie instead of strudel, and raise their children according to the American Dream. They considered themselves incredibly lucky, especially those who had once been card-carrying Nazis. Their adopted country valued their services, gave them healthy funding, and seemed eager to forget their past.

Eventually, von Braun's rockets would outgrow the facility at Huntsville. Building them was one thing, firing them another. Anticipating this development, the government established, in 1949, a Joint Long Range Proving Ground along the Banana River in Cape Canaveral, Florida. Booster rockets developed and built at Huntsville were sent in pieces on ocean-going barges to the Cape Canaveral facility, where they were assembled and fired. The separation of construction and testing had the effect of creating two boomtowns. Over the next twenty years, Huntsville's population would increase tenfold. Cape Canaveral's growth rate was even steeper.

One era ended as another began. The first launch at Cape Canaveral, involving a two-stage Bumper 8 rocket, took place on July 24, 1950. The first stage was a V-2 rocket made in Germany, and one of the last in the American booty of war. It was supposed to travel on a near horizontal trajectory, as suited the flight of a nuclear missile. The first stage performed well, but at the point where separation was supposed to occur, a malfunction threw the rocket into a dangerously wild course. It was destroyed by remote control. Henceforth, American missiles would be made in America.

A few months later, RAND returned to the subject of satellites. Influenced by the successful Soviet atomic test, analysts concluded that a satellite would be a "novel and unconventional instrument of reconnaissance." Rarely has so much significance been packed into six words. The operative word was "reconnaissance"—the ability to see the world from up high had implications even the imaginative people at RAND couldn't quite get their heads around. Possibilities ranged from the benign—meteorological prediction—to the sinister, spying. Finally, there was that word "unconventional." The unusual nature of this technology raised enormous questions, among them that of legality. Was it lawful to spy on the Soviets from outer space, if it wasn't legal to fly over their territory in an airplane? How would the enemy react? Would the launch of a satellite worsen relations with the Kremlin? "Our objective," RAND concluded, "is to reduce the effectiveness of any Soviet counteraction that might interfere with the satellite reconnaissance operation before significant intelligence results are secured. Perhaps the best way to minimize the risk of countermeasures would be to launch an 'experimental' satellite on an equatorial orbit."[11] In other words, since such a satellite would not fly over Soviet territory, it could be used to test the issue of freedom of space and perhaps serve as a precedent when satellites of more sinister intent were launched.

All the musing about satellites started from the assumption that the Americans would be first in space. RAND thought that the feat would delight America's allies, and frighten her enemies. "We may assume that the Soviet leadership will consider the actual or possible use of novel reconnaissance facilities against it as a major threat," a 1950 report predicted. "As long as there are uncertainties about the capabilities of the satellite, a general state of anxiety and frustration will probably prevail."[12] That assessment was astute, but RAND wrongly predicted which nation would suffer the anxiety.

RAND was thinking about what could be done with satellites in the sky long before the military or the government felt enthusiastic about putting them there. In contrast, in the Soviet Union, the eagerness to put an object into orbit was great, but surprisingly little discussion took place about the utility of doing so. This had enormous implications for the space race to come: the Russians may have been first in space, but when the United States finally got around to joining them, they actually understood what satellites could do.

As an indication of things to come, a V-2 launch in June 1948 was hugely important. The rocket was called Albert, in honor of its passenger, a chimp stuffed into its nose cone. The test was supposed to determine whether a primate could withstand the g-forces of lift-off, the effect of which would multiply his weight by a factor of six. After being shot into the sky, Albert's little capsule was supposed to drift peacefully back to Earth on a parachute, whereupon the chimp could go back to eating bananas. As it turned out, the g-forces proved immaterial since Albert died of suffocation, the first martyr of the U.S. space program. A year later, Albert II (rocket scientists did not waste time thinking up names) was launched with another unwilling primate aboard. This one fared better, at least as far as his oxygen supply was concerned, but died when its return to Earth proved bumpier than expected. Two more attempts that year suggested that primates were quite good at withstanding g-forces, but both passengers died while awaiting rescue from their capsules.

The Albert tests demonstrated that von Braun and his team had in mind not just space exploration (which could be achieved with cameras, radar devices, and the like) but also space travel. They also revealed that adding a living organism to the rocket equation made it infinitely more complicated. But that complexity was what drove von Braun. He understood that the exploration of space could never really get off the launch pad unless it was rooted in the familiar. Merely shooting instruments into the sky would always have limited public appeal. Americans, if they thought about space at all, thought Buck Rogers. The only type of project considered worthy of their support was one that would allow them to live vicariously the comic book life.

In 1947, von Braun completed *The Mars Project,* a book which, as the name implied, speculated on the shape of a future manned mission to the Red Planet. Though the space craze had yet to begin in earnest, von Braun had an uncanny sense of the developing mood. He also knew how to nurture that mood. He was an enthusiastic salesman eager to promote space exploration to any group willing to give him a platform. While he was quite happy to take the military's money, his ambitions aimed higher, as he explained in 1952:

Many a serious rocket engineer, while firmly believing in the ultimate possibility of manned flight into outer space, is confident that space flight will somehow be the automatic result of all the efforts presently concentrated on the development of guided missiles and supersonic airplanes. I do not share this optimism. . . . The ultimate conquest of space by man himself is a task of too great a magnitude ever to be a mere by-product of some other work.[13]

Though a massive intellect, von Braun had the instincts of a populist; he knew precisely how to tailor his message to his audience and felt no aversion about sharing complex thoughts with mere plebeians. He was helped by a number of other talented space enthusiasts who knew how to talk to the people, among them Arthur C. Clarke, whose 1951 *Exploration of Space* set out in confident terms the road (or flight path) ahead. It was a huge best-seller and a Book of the Month Club selection. In addition, Willy Ley's *Rockets, Missiles and Space Travel*, first published in 1944, was massively popular, remaining constantly in print. That book was followed in 1949 by *The Conquest of Space*, lavishly illustrated by the premier space artist of the time, Chesley Bonestell. It explained with admirable simplicity how space might be conquered, and what the first explorers might find "out there." The consistent message in these books was that space exploration would be difficult, dangerous, and expensive, but also, in a way, inevitable. As Ley wrote:

It is the story of a great idea, a great dream, if you wish, which probably began many centuries ago on the islands off the coast of Greece. It has been dreamt again and again ever since, on meadows under a starry sky, behind the eyepieces of large telescopes in quiet observatories on top of a mountain in the Arizona desert or in the wooded hills near the European capitals. . . . It is the story of the idea that we possibly could, and if so should, break away from our planet and go exploring to others, just as thousands of years ago men broke away from their islands and went exploring to other coasts.[14]

Or, as Clarke prophesied:

Even if we never reach the stars by our own efforts, in the millions of years that lie ahead it is almost certain that the stars will come to us.

Isolationism is neither a practical policy on the national or the cosmic scale. And when the first contact is made, one would like to think that Mankind played an active and not merely a passive role—that we were the discoverers, not the discovered.[15]

The complexity of the task was not seen as an obstacle but rather as an indicator of its importance. Progress, the message went, implied tackling difficult challenges before their time.

The turning point in the marketing of the rocket age came in 1952 when *Collier's* magazine ran an eight-part series on space travel. The lead article of the first installment (entitled "Man Will Conquer Space Soon") was by von Braun, who weaved a fantastic tale about a massive Earth-orbiting space station that would serve as a docking port for cosmic taxis. According to the dream, its constant rotation would generate artificial gravity, making it livable for an extended period. The station would serve as a foundation for lunar expeditions, and eventually for trips to Mars. According to von Braun's grand plan (suitably glamorized for *Collier's* readership), a fleet of ten vehicles would be built in space before embarking like a cosmic fleet for the Red Planet. Once there, a manned landing boat would be dispatched, setting down with skis on the planet's polar ice cap.

The predominant theme in the *Collier's* series was that the rocket age had already started, and that, at the very least, a lunar landing would occur "in our lifetime." Space exploration was presented as an inevitable stage in man's evolution. The series brought cutting-edge science to the people by using experts who were not afraid to communicate. The popular historian Cornelius Ryan, a senior contributor at the magazine, helped the scientists provide drama and punch, while Bonestell and Fred Freeman sparked the imagination further with vivid images of what space vehicles would look like. The magazine's message was simple: "Man will conquer space soon. What are we waiting for?"[16]

The series was not all about romance and adventure. Woven into the comic-book story was a message about political expediency, designed for those who were otherwise unenthusiastic about rockets. An editorial claimed that Americans had to "immediately embark on a long-range development program to secure for the West 'space superiority.' If we do not, somebody else will. That somebody else very probably would be the Soviet Union." Complacency was the greatest danger:

A ruthless foe established on a space station could actually subjugate the peoples of the world. Sweeping around the Earth in a fixed orbit, like a second Moon, this man-made island in the heavens could be used as a platform from which to launch guided missiles. Armed with atomic warheads, radar controlled projectiles could be aimed at any target on the earth's surface with devastating accuracy.[17]

This was all complete bunk, but fear is an effective persuader when romance proves inadequate.

In his *Collier's* piece, von Braun boasted that a Moon mission would take place within the next 25 years. "There are no problems involved to which we don't have the answers," he claimed, "or the ability to find them—right now."[18] Few industry experts shared this view. A 1953 poll found that only one quarter of those involved in the space industry believed that man would land on the Moon by 1969. Most thought that such an event would not occur until the twenty-first century, if at all.[19] *Time* magazine, eager to be seen as a serious newsweekly, poured scorn on *Collier's*, calling the space series a happy mix of fact and fiction. The public, *Time* regretted, had been oversold the idea of space travel, to the point where it "apparently believes that spaceflight is just around the corner."[20] Von Braun devoted himself to defeating skepticism of this sort. As a result of the series, he became a household name, which suited his ego perfectly. He called for a "daring, inspiring program that has a real chance of controlling the world."[21]

As proof of this popularity, in 1954 Walt Disney recruited von Braun to be the brains behind a television series on space travel being produced for ABC. The first program, "Man in Space," aired on March 9, 1955, and was rebroadcast twice during the following summer.* The programs were a launch pad for the theme park Disney was building in southern California. In the park, Tomorrowland gave popular form to von Braun's ideas. Disney provided rocket ships, simulated trips to the Moon, People Movers, and, circling the park, a monorail which looked like something space colonists would someday use to get to work. "After entering the . . . space port," Disney explained, "visitors may experience the thrills that space travellers of the future will encounter when rocket trips to the Moon become a daily routine."[22] No one

* Two other programs were titled "Man and the Moon" and "Mars and Beyond."

seemed to care that the rockets in the park looked disturbingly like V-2s or that the Autopia had echoes of Hitler's Autobahn.

At Disneyland, the futuristic Tomorrowland was just a gentle stroll from Frontierland. One era morphed smoothly into another, with the time traveler picking up cotton candy, a hot dog or a Mouseketeers hat along the way. The values embodied in coonskin caps, log cabins, and muskets were perfectly suited to that of rocket ships, space stations, and people movers. The common theme was exploration, enhanced by entrepreneurship. The West was conquered by the horse and Winchester rifle, and space would be subdued by the rocket and ray gun. Close on the heels of Buck Rogers, bankers and bureaucrats would follow.

Americans had grown up on the ideas of Frederick Jackson Turner, even if they'd never heard of him or read his books. Many instinctively agreed with Turner's view that American progress was governed by the frontier imperative, and that the yearning to colonize the continent provided a vitality which distinguished the United States from other nations. When the frontier was officially closed, that vitality came under threat—or so it was thought. In truth, the Turner thesis was founded upon a romanticized vision of the frontier which bore little relation to reality. The American West was not a pristine democratic paradise in which pioneers automatically acted nobly and avarice was suspended. Settlers brought the imperfections of society with them in the holds of their Conestoga wagons. The pioneering life seemed idyllic to everyone except the pioneer.

Nevertheless, for those religiously devoted to Turnerism, space seemed a natural outlet for the pent-up American spirit. It was readily accepted as "the final frontier," as if there were no challenges left to tackle on Earth. Conquering space resonated perfectly with myths about conquering the American West—a new generation would be able to share the experiences of their forebears. First would come the brave explorers, who would ride rockets to new outposts of the American Dream. Then would come the intrepid pioneers establishing colonies on distant planets. Then would come commerce and, rather like Alaska, the Moon and Mars would eventually become states of the Union.

Missing from this well-scripted scenario was a sense of reality and a willingness to ask the simple question: "Why?" That question did not remotely bother von Braun, since he was more interested in "When?" He understood that the essential prerequisite to a fully fledged space program was a public sufficiently excited by the idea of space travel to pay for it with their tax dollars. One way to achieve this, he understood, was through cheap gimmickry. In 1954, a book by Jonathan Norton Leonard entitled *Flight into Space* ended with an appeal by von Braun for the creation of an artificial star which would have no other purpose than that of wowing the people. He had in mind a "small rocket," of minimal cost, which would "shine at dusk against a dark sky. Once established in orbit, it will inflate a white plastic balloon 100 feet in diameter. Swinging swiftly around the Earth 200 miles above the surface, the balloon will gleam as brightly in the sunlight as a first magnitude star." The satellite would do precisely nothing, but von Braun insisted that it would "make an enormous impression on the people of Asia."[23]

Attempts to sell space worked brilliantly. In 1955, as a result of the *Collier's* series, the Disney programs, countless science fiction films, and hundreds of speeches by von Braun, Americans were beginning to believe in space travel. According to one poll, 38 percent believed that Americans would reach the Moon "in the next fifty years"—a rise of 23 percent from when the same question was asked six years earlier.[24]

In March 1955, von Braun and 102 of his fellow "prisoners of peace," in addition to their families, were all officially made American citizens in a mass ceremony in Huntsville, Alabama. The ceremony bore a remarkable similarity to the mass baptisms popular within southern evangelical movements. The past was forgotten and sin washed away.

Three years later, and a full thirteen years after the war, von Braun was invited to give the commencement address at St. Louis University. He chose as his theme morality and science, telling the audience gathered in the June sunshine:

> Technology and ethics are sisters. While technology controls the forces of nature around us, ethics controls the forces of Nature within us. . . . I think it is a fair assumption that the Ten Commandments are entirely adequate, without amendments, to cope with all the problems the Technological Revolution not only has brought up, but will bring up in the future. . . .
>
> It has frequently been stated that scientific enlightenment and religious belief are incompatible. I consider it one of the greatest tragedies of our times that this equally stupid and dangerous error is so widely believed.[25]

That came from the man who used slave labor to build rockets.

The American people, keen to live in the present, willingly ignored von Braun's past. All that mattered was that the silver-tongued German would take them to the Moon.

4

Sputnik

Within American politics, those to the left of Robert Heinlein numbered nearly two hundred million, while those to the right could hardly fill a bus. The science fiction novelist didn't think much of the way President Harry Truman was looking after the security of the United States. The nation, he thought, had a clear-cut and dangerous enemy, but was doing nothing substantial to confront it. Eager that Americans should wake up to the dangers threatening them, he decided to weave his political message into his fiction. The result was the film *Destination Moon* (1951), on which Heinlein collaborated with the director George Pal. In the film, a group of politically astute businessmen, fed up with the timid foreign policy of their government, decide to sponsor a private venture to travel to, land on, and colonize the Moon. At one point a retired general explains the project to a group of businessmen:

> We are not the only ones who know the Moon can be reached. We're not the only ones who are planning to go there. The race is on and we'd better win it because there is absolutely no way to stop an attack from outer space. The first country that can use the Moon for the launching of missiles will control the Earth. That, gentlemen, is the most important military fact of this century.[1]

Actually, it wasn't a fact at all. But fiction is often more exciting than the truth.

Truman struggled with the problem of maintaining American security while keeping the federal budget under control. His solution was nuclear weapons, a cheaper and more dependable alternative than a massive land army. His successor, Dwight Eisenhower, was equally frugal. "This country can choke itself to death piling up military expenditures just as surely as it can defeat itself by not spending enough for protection," he repeatedly argued.[2] Unfortunately, the safety afforded by nuclear weapons was less dependable after the Soviets tested their first device in 1949. What Eisenhower desperately wanted to resist was a crippling arms race, which would not necessarily make America safer, but would definitely make her poorer. In a 1953 speech to the Society of Newspaper Editors, he referred eloquently to the dilemma facing the world:

> Every gun that is made, every warship launched, every rocket fired signifies, in the final sense, a theft from those who hunger and are not fed, those who are cold and are not clothed.
> This world in arms is not spending money alone.
> It is spending the sweat of its laborers, the genius of its scientists, the hopes of its children.
> We pay for a single fighter plane with half a million bushels of wheat.
> We pay for a single destroyer with new homes that could have housed more than 8,000 people. . . .
> This is not a way of life at all, in any true sense. Under the cloud of threatening war, it is humanity hanging from a cross of iron.[3]

The solution to the dilemma, Eisenhower thought, might be found in the sky. Satellites, by offering the possibility of knowing precisely what the Russians were doing, made it possible to achieve the same level of protection with a much smaller military force. That, at least, was how things seemed in 1953.

But within the Eisenhower solution there lurked a quandary. While satellites might make it possible to avoid a financially crippling arms race, they still implied a costly commitment of their own. Whenever anyone spoke of the urgency of the satellite issue, references to the Manhattan Project were made. Yet that initiative had cost the United States $2 billion and had resulted in a government-sponsored industrial conglomerate bigger than the auto industry. Eisenhower understood that satellites would again require the government to play a huge role as sponsor and manager, something he abhorred, both for reasons of economy and ideology. Nor was there any guarantee that the advent of satellites would make the Cold War antagonists feel more secure. Acrimony might simply be extended into outer space.

The complexity of the satellite issue perhaps explains why the government seemed to blow hot and cold. Eisenhower wanted to dampen public enthusiasm; he did not want to race into space simply to satisfy popular emotion. On November 17, 1954, therefore, the secretary of defense denied that the United States was working on a satellite and further stated that he would not be surprised if the Soviets were first into orbit. Avoiding a populist clamor was, however, difficult when the ubiquitous Wernher von Braun was talking up space from every podium. His most recent project was the unfortunately named Slug (later called Orbiter), a five-pound midget which, he claimed, could be put into orbit for virtually nothing. It would rely on existing technology, with the army supplying the booster and the navy the tracking facilities. Ever the salesman, von Braun argued that "a man-made satellite, no matter how humble . . . would be a scientific achievement of tremendous impact." But, he warned, since it would be easy for the United States to put such a device into orbit, "it is only logical to assume that other countries could do the same." And, he stressed, "It would be a blow to U.S. prestige if we did not go first."[4]

Satellites were given an additional boost with the advent of the International Geophysical Year (IGY), a global think-fest aimed at unlocking the deepest secrets of the universe. The idea was born at a party hosted by the astrophysicist James Van Allen on April 5, 1950. A group of eminent scientists had gathered to meet the renowned geophysicist Sydney Chapman, who was in Washington on a brief visit from Britain. Talk turned to the course the group wanted their research to take, and particularly to the attractiveness of measurements and experimentation performed in the upper atmosphere. Lloyd Berkner, recently appointed to head the new Brookhaven National Laboratory, suggested organizing another International Polar Year. The first such event had been held in 1882, the second in 1932. The year-long symposium was designed to bring together scientists from all over the world to cooperate on revolutionary projects. Logic suggested that the next event would be held in 1982, but the scientists preferred 1957–58, for a number of reasons. The first and most practical was the fact that a period of intense solar activity would occur then, offering unique opportunities for discovery. They also realized that technology had accelerated since 1932, rendering a fifty-year gap nonsensical. It seemed wise to wed new technology to a coordinated effort at research. Finally, the organizers felt a sentimental yearning to re-create the worldwide community of science that had been destroyed by the Second World War. The spirit of cooperation would provide a welcome contrast to the antagonism of the Cold War.

Berkner's enthusiasm proved wildly contagious. Before long, committees were formed, resolutions passed, and schedules formulated. The steering group agreed that the research would not be restricted to the polar regions and that, in line with this broadening of scope, the event would be called the International Geophysical Year. The boldest plan called for the IGY to be marked by one epoch-making feat that would draw worldwide attention and push research to an entirely new dimension. They had in mind the launch of the world's first satellite. It is important to bear in mind that the IGY was intended to be a scientific event of no specific national identity. It mattered not which nation launched the satellite since, as the organizers naively believed, the achievement would be seen as a leap forward for science, not for any specific country. In order to maximize the potential (and increase the chances of success), the year in question was defined as eighteen months—from July 1, 1957 to December 31, 1958.

The IGY plan harmonized perfectly with Eisenhower's subtle approach. On February 14, 1955, the Technological Capabilities Panel, headed by James Killian of the Massachusetts Institute of Technology, advised the government that it would be a good idea to launch an overtly scientific satellite in order to set a precedent for the principle of "freedom of space." A similar suggestion, it will be recalled, had been mooted by RAND a few years before. The panel, originally commissioned to investigate the possibility of a surprise nuclear attack against the United States, concluded that reconnaissance satellites would significantly reduce the likelihood of such an attack. But, since unauthorized overflights of a country by airplanes were illegal, a different principle governing satellites would have to be established before they could prove their worth. The panel felt that if the first satellite was scientific, a precedent would be set, and future spy satellites could fly through that legal loophole.

This proposal was formally (but not publicly) endorsed by Eisenhower. The choice of a satellite program was given to the assistant Secretary of Defense for Research and Development, Donald Quarles. On May 20, 1955, his office produced NSC-5520, which argued in favor of a quick leap into space. "From a military standpoint," the document maintained, "intelligence applications strongly warrant the construction of a large surveillance satellite. While a small scientific satellite cannot carry surveillance equipment and therefore will have no direct intelligence potential, it does represent a technological step toward the achievement of the large surveillance satellite, and will be helpful to this end." In addition, a small satellite, launched before the Russians, would have enormous symbolic importance:

> Considerable prestige and psychological benefits will accrue to the nation which is successful in launching the first satellite. The inference of such a demonstration of advanced technology and its unmistakable relationship to inter-continental ballistic missile technology might have important repercussions on the political determination of free world countries to resist Communist threats. . . . Furthermore, a small scientific satellite will provide a test of the principle of the Freedom of Space.[5]

The "freedom of space" argument was reiterated a month later in NSC 5522, which forcefully advocated taking advantage of "an early opportunity to establish a precedent for distinguishing between 'national air' and 'international space,' a distinction which could be to our advantage at some future date when we might employ larger satellites for surveillance purposes."[6]

Technically speaking, it did not actually matter which nation established the precedent of "freedom of space." There was, however, the issue of prestige. NSC-5520 recommended that the government throw its weight behind an American effort under the auspices of the IGY. "The stake of prestige that is involved makes this a race that we cannot afford to lose."[7] "Prestige" had quickly become the favorite word among those pushing for an ambitious space program. Given the military stalemate of the Cold War, prestige had replaced power in the calculation of a nation's strength.

NSC-5520 quickly became administration gospel on the subject of satellites. By July, Quarles had on his desk three proposals: the Army Ballistic Missile Agency's Orbiter, the Naval Research Laboratory's Viking, and a makeshift air force proposal based on modifying an Atlas rocket. Quarles left the decision to a committee made up of two representatives from each of the three services, in addition to two nonattached delegates selected by him. The committee was headed by Homer Stewart of the Jet Propulsion Laboratory (JPL).

The NRL's proposal was the most coherent and logical of the three, given the requirement that the satellite be overtly scientific. The satellite, called Vanguard, was designed to stuff as much scientific instrumentation into as small a package as possible. As such, it harmonized perfectly with the aims of the IGY, since it was so benignly scientific. The other two proposals, in contrast, were makeshift adaptations of military hardware motivated by the desire to get an object—any object—into orbit as soon as possible.

While Eisenhower was preparing for the satellite age, he simultaneously sought ways to increase trust between the United States and the Soviet Union. He wanted desperately to establish a climate that would allow both sides to avoid the rampant proliferation of nuclear weapons. The problem arose, he felt, because of ignorance and distrust. With neither side knowing what the other was doing, both took what seemed the safest course, namely to accumulate as many weapons as possible. Eisenhower wanted to break this vicious circle. To this end, he proposed his "Open Skies" initiative on July 21, 1955, under which the two belligerent nations would keep each other fully informed of their capabilities, which could be verified by unrestricted aerial reconnaissance. "I only wish that God would give me some means of convincing you of our sincerity and loyalty in making this proposal," Eisenhower told the Soviet delegation at Geneva.[8]

Open Skies was a genuinely noble idea, but one laced with cynicism. The United States had nuclear superiority but wanted a clearer idea of how many weapons the Soviets had and where they were located. Since secrecy was an advantage for the Soviets, given that it allowed them to bluff, they had no practical reason to accept the proposal. But, by rejecting it, they made it seem like they did not genuinely want peace. The proposal also allowed the United States to go ahead with a full program of reconnaissance, on the presumption that overflights were necessary in order to make Open Skies work. Spying suddenly became a noble instrument of peace.

Meanwhile, on July 28, just a week after Open Skies was proposed, James Hagerty, the White House press secretary, formally revealed what had been decided long before, namely that the United States would launch a small satellite as part of the IGY. In line with the RAND and NSC recommendations, the first satellites would be innocent ones designed purely for scientific research.

On August 24, the Stewart committee opted in favor of the NRL's Viking/Vanguard proposal. While the rocket would be based on existing Viking technology, it would be an entirely separate program from the one the navy was already pursuing. This was in line with Eisenhower's wishes that the IGY satellite should not be carried aloft by a rocket already part of an existing weapons program. Such a move would underline the separation between scientific and military ventures, but also make certain that defense programs were not sidetracked by the IGY initiative. Orbiter, poorly conceived as a scientific device, failed to satisfy these criteria. Since the Redstone rocket was well known to be a modified V-2, Orbiter could not escape its reputation as a weapon of destruction, and a German one at that. But, since Redstone was a much more promising rocket than Viking, this implied that leading the space race was less important to Eisenhower than establishing the civilian character of the U.S. effort and perhaps setting a precedent for "freedom of space." Eisenhower had desperately tried to find a solution that satisfied his need for a satellite reconnaissance system, without sparking a new arms race and unduly alarming the American people. His was a rational, sensible, honorable, and ultimately futile policy.

It was generally agreed that space fantasies could not be allowed to interfere with the nation's defense. Meeting in June 1955, The ICBM Scientific Advisory Committee "unanimously agreed that any Satellite program, Scientific or Reconnaissance, which is dependent upon components being developed under the ICBM program, would interfere with the earliest attainment of an ICBM operational capability."[9] Charles Lindbergh, who served on that committee, recalled one meeting where the subject of an orbiting satellite was broached:

> Everyone present agreed that a satellite-launching project was practical, important and deserving of support. But what priority should be assigned to it? We were working under staggering concepts of nuclear warfare, and the imperative need to prevent a strike against the United States by the Soviet Union. Our mission was to speed the development of intercontinental ballistic missiles with sufficient retaliatory power to discourage attack. Members of the committee believed that the security of our nation and civilization would depend on our ability to shoot across oceans quickly and accurately by the time Russia had missiles available for shooting at us. A shortage of scientists, engineers and facilities existed in all fields of missiles and space. Military projects needed every man available. The consensus of committee opinion was that we should concentrate on security requirements and assign to the Vanguard program a secondary place. There would be time to

orbit satellites after our nuclear-warhead missiles were perfected and
adequate marksmanship achieved.[10]

This advice was perfectly in line with what Eisenhower himself felt. It
is ironic that, by making a responsible decision in the long-term interests of the nation's defense, the president opened himself to criticism
that he was neglecting the nation's safety. The ability to put a small ball
into space had become confused with national security. Eisenhower had
decided that the most important race was that involving ICBMs. He
won that race, but never received credit for his victory.

The decision to go with the navy proposal implied a cancellation of
the other two. On September 9, both the army and the air force were
strictly instructed to cease all satellite work. This suggests either a
supreme confidence that Vanguard would work, or a lack of concern for
winning the race. Had Eisenhower been consumed by the need to beat
the Russians, the sensible thing would have been to permit all three
projects to go ahead, the logic being that one would surely succeed. Instead, all eggs were placed in the navy basket. An indication of the administration's thinking can be garnered from the fact that on September
13 the army was instructed to concentrate its energies on the development of the Jupiter intermediate range ballistic missile, a four-stage version of the Redstone. This project, headed by Killian, was expected to
launch by September 1956. While it did not carry the cachet of an orbiting satellite, it was hugely important to Eisenhower, much more so than
Vanguard. While von Braun and his army boss, Major General John
Medaris, felt unfairly treated, they had in fact been given what the president felt was the most important job.

Von Braun and Medaris did not, however, take the decision lightly.
They launched a stinging attack on Vanguard, arguing that it was incapable of fulfilling the promises the NRL had made. Jupiter, they maintained, was the only project with a realistic hope of succeeding. That argument was bolstered when a test on September 20, 1956, stunned the
American rocket community. The first three Jupiter stages performed
perfectly, taking the rocket to an altitude of 682 miles and a speed of
13,000 miles per hour, just short of the velocity necessary to achieve
orbit. This achievement was all the more impressive given that the
fourth stage had, under the instructions of the DoD, been filled with
sand, in order to make certain that it could not reach orbit. Von Braun

and Medaris rightly felt that their toy could easily win the race into space.

Eisenhower, however, had no desire to win a race he claimed did not exist. Reflecting on the matter in 1965, he remarked:

> Under no circumstances did we want to make the thing a competition, because a race always implies urgency and spectacular progress regardless of cost. . . . Neither then nor since have I ever agreed that it was wise to base any of these projects on an openly announced competition with any other country. This kind of thing is unnecessary, wasteful and violates the basic tenets of common sense.[11]

His strategy was subtle, but the time for subtlety was quickly running out. In any case, lovers of powerful rockets are not very sensitive to nuance. Thanks in part to the shenanigans of von Braun and Medaris, the coherence of the American rocket venture had begun to unravel. According to Arthur C. Clarke, on one occasion von Braun was preparing to "accidentally" launch a satellite by allowing a Jupiter rocket to perform to its maximum potential. In the nick of time, the DoD got wind of the intention and put a stop to it. A member of von Braun's team claimed in 1998 that Eisenhower had installed his observers at ABMA, so that the rocketmeister could be kept in line. While both these claims might be technically untrue, both have a basis in fact. There is no doubt that the ABMA team was pursuing its own egotistical fantasies in opposition to the wishes of the government.

On November 26, 1956, ABMA's wings were clipped further. In an attempt to rationalize missile programs, the DoD announced that in the future the navy would, appropriately, have control of missiles deployed from ships. The air force, in keeping with its coveted role as the overseers of the nuclear deterrent, would manage the ICBMs, and the army would be confined to tactical missiles with a range of less than 200 miles. While ABMA would continue to oversee the development of the Jupiter missile, it would henceforth be an exclusively air force weapon. The decision proved devastating for morale at Huntsville.

Medaris, in direct contravention of orders, decided to hold on to components of the Jupiter rockets, rather than turn all of them over to the air force. He had a hunch that Vanguard would fail and that a time would come when the country would desperately need something to shoot into space. In anticipation of such an event, he spirited away the hardware necessary to put a device into orbit within four months of being asked.

In late July 1955, Hagerty announced the U.S. determination to launch a satellite during IGY. The Soviets responded by saying that they, too, would put an object in orbit. On August 2, Leonid Sedov upped the ante even further with a rather oblique suggestion that the Soviets would be able to launch something "of larger dimensions" than that which Hagerty had in mind.[12]

In early 1956, *National Geographic* revealed that the first man-made satellite, to be called Vanguard, was being built by the U.S. Navy and would be launched during the IGY. In May 1957, a cheap book for space enthusiasts called *Discover the Stars* featured a picture of Vanguard on its cover and enclosed plans for hobbyists to build a model of the satellite. The book forecast that it would be launched from Cape Canaveral in Florida in early 1958. In fact, that prediction seemed a bit rash since the rocket designed to carry the satellite aloft was proving highly temperamental. Nevertheless, the fact that it was about nine months behind schedule hardly seemed to matter, since most people assumed that, whatever happened, it would still be the world's first satellite.

Americans slept soundly under a blanket of complacency. On September 27, 1957, Ernst Stuhlinger, an engineer at ABMA, urged Medaris to put pressure on the government. "A Russian satellite will soon be in orbit," he warned. "Wouldn't you try once more to ask the Secretary [of Defense] to go ahead with our satellite? The shock for our country would be tremendous if they were the first in space!" Medaris tried to calm Stuhlinger. "Now look . . . don't get tense. You know how complicated it is to launch a satellite. These people will not be able to do it. . . . Go back to your laboratory and relax!"[13] On October 3 the aerospace engineer Max Faget was trying to convince John Crowley, director of research at the Washington office of the National Advisory Committee for Aeronautics (NACA), of the importance of developing a four-stage, solid propellant launch vehicle. Crowley, however, wasn't interested. In frustration, Faget shouted: "What if the Russians launch a satellite tomorrow?"[14]

They did precisely that.

. . .

Nikita Khrushchev was stunned when, upon assuming the premiership after Stalin's death in 1953, he learned how much progress had been made by Sergei Korolev:

> I don't want to exaggerate, but I'd say we gawked at what he showed us as if we were sheep seeing a new gate for the first time. When he showed us one of his rockets, we thought it looked like nothing but a huge, cigar-shaped tube, and we didn't believe it would fly. Korolev took us on a tour of the launching pad and tried to explain to us how a rocket worked. We were like peasants in a marketplace. . . . We had absolute confidence in comrade Korolev. When he expounded his ideas, you could see his passion burning in his eyes, and his reports were always models of clarity. He had unlimited energy and determination, and he was a brilliant organizer.[15]

For Korolev, the IGY seemed a perfect opportunity to show off the power of his R-7 rocket, which officially went into production in 1953. If Viking was a stiletto, the R-7 was a bludgeon. The rocket was an ugly brute that sat on its launch pad with menace steaming from its vents. It didn't pierce the atmosphere, it assaulted it. Twenty separate engines burned kerosene and liquid oxygen at a prodigious rate, producing 1.1 million pounds of thrust. This was essential since the rocket was born in an era when Russian atomic weapons were crude and heavy. The Russian space agency was reaping the benefits of the bomb designers' failure to miniaturize. Since it was designed to carry a warhead weighing five tons on an intercontinental ballistic trajectory, lifting a small object into orbit was child's play. In fact, the rocket was technically powerful enough to lift a 1,500-kilo payload into orbit, and, as such, was four times more powerful than Redstone.

The first six tests were dramatic failures. Success came finally on August 3, 1957. Just over three weeks later, Korolev got approval for a satellite launch. Having observed ABMA's progress as closely as espionage could allow, he assumed that the September 20 launch had indeed been an attempt to reach space. Since he was in no doubt that a race had started, he redoubled efforts to get his rockets ready. His original intention was to push the R-7 to its limits by launching a 1,500-kilo orbital laboratory capable of a wide range of experiments. But then, on hearing that Vanguard was progressing, and frightened by what von

Braun might do with Redstone, he decided to take the safer approach of launching a smaller satellite early in the IGY, with the intention that the bigger one would be launched in early 1958.

Though the Soviets were renowned for secrecy when it came to weaponry, they loved to boast about their rockets. Anyone inclined to pay attention would have noticed an obsessive Russian need to discuss rocketry during the period 1951 to 1957. Their claims were widely reported in the American press, with twenty-five articles on the subject in the *New York Times*. One article reported (very accurately, as it turned out) that the first Soviet satellite would be 18 inches in diameter and would weigh about 100 pounds. Then, on September 17, the centennial of Tsiolkovsky's birth, came the boldest statement: a satellite launch was coming "soon," the Soviets announced. A further hint had come at the Comité Spécial de l'Année Géophysique Internationale, a conference at the National Academy of Sciences in Washington on September 30. Representatives from five countries attended to hear their colleagues lay out plans for space exploration and to agree upon guidelines for sharing of findings. Sergei Poloskov, a member of the Soviet delegation, unveiled his country's plans for a satellite called Sputnik (he pronounced it "Spootnik") due to be launched "imminently." He claimed that the satellite would broadcast alternately on frequencies of 20 and 40 megacycles, in contradiction of the IGY approved frequency of 108 megacycles.* American representative, Homer Newell, complained that the Vanguard tracking devices, which would be up and running the following day, were set up according to the IGY-agreed frequency.[16] Argument over the broadcast frequency seems to have overshadowed the import of what Poloskov was saying.

On August 31, 1957, Drew Pearson, the noted muckraker, used his "Washington Merry-Go-Round" column to predict that the Soviets would launch a satellite ahead of the Americans.[17] But warnings like this went largely unheeded, because most Americans simply could not believe that any other country, especially not a communist one, would beat them in a contest of technology. The fact that the Soviets had built an atom bomb much quicker than anyone predicted was treated as the exception that proved the rule, since their atom bomb, it was smugly noted, was built from stolen American plans.

---

* In order to ensure that as many people as possible would hear the broadcast, Korolev decided to set the frequencies between two popular ham radio bands.

.   Those paid to know what the Russians were doing had a reasonably accurate conception of the state of affairs. A prediction put forth by
RAND early in 1957 set a Russian launch date that was only two weeks
wide of the mark. The CIA, according to its one-time director Allen
Dulles, made a similar prediction. But these conclusions could not be
discussed in the open because the evidence on which they were based
had been obtained by methods that supposedly did not exist. For instance, in June 1957, a U-2 spy plane accidentally strayed over the Tyuratam rocket test site in Kazakhstan and returned with enough evidence
to allow intelligence analysts to make reasonably accurate guesses
about Soviet capabilities.

Eisenhower would obviously have been privy to this information,
but he stubbornly refused to panic. He accepted that Redstone could
have put a missile in orbit before the IGY. But to do so with a military
rocket, and without the scientific justification the IGY provided, would
send the wrong signals to the Soviets. The president was playing a very
canny game. If the Soviets were first to put a satellite in orbit, and if that
launch occurred during the IGY, then the precedent for freedom of
space might be established. Eisenhower understood that satellites
would eventually prove useful for spying, and wanted nothing to get in
the way of exploiting them for that purpose. The greater good of freedom of space was, therefore, much more important to him than that of
beating the Russians. Far from being worried by the possibility of the
Russians being first in space, Eisenhower actually welcomed that scenario.

His frame of mind harmonized nicely with that of the scientists
participating in the IGY. But satellites went into space aboard rockets,
and rockets were vehicles for nuclear warheads. Eisenhower might
have been able to achieve, partly by stealth, the freedom of space goal
he sought, but he could not control the political fallout of a Soviet satellite being first in orbit. He could not prevent the launch of the first satellite being turned into a massive propaganda coup. By not worrying
about the possibility that the Russians might be first in space, he gifted
them a huge political victory.

The Soviets fully appreciated the importance of such an event.
Though they, too, operated under the cover of the IGY (which partly explains their extraordinary openness), they also understood fully that
the launch would be hugely important politically. They had almost
nothing to lose and everything to gain from a launch. A failure would

not reflect badly on them, since they had no technological reputation to protect. In any case, anyone familiar with rocketry understood that failures were expected. If, however, they succeeded, the effect would be monumental.

At 1912, Greenwich meridian time on October 4, the R-7 monster came to life at the Baikonur complex.

> The clear tones of a bugle were heard above the noise of the machines on the pad. Blinding flames swirled about, and a deep rolling thunder was heard. The silvery rocket was instantly enveloped in clouds of vapor. Its glittering, shapely body seemed to quiver and slowly rise up from the launch pad. A raging flame burst forth and its candle dispelled the darkness of night on the steppe. So fierce was the glare that silhouettes of the work towers, machines, and people were clearly outlined. . . .
>
> "She's off! Our baby is off!" People embraced, kissed, waved their arms excitedly and sang. Someone began to dance, while all the others kept shouting, "She's off! Our baby is off!"[18]

A lot went wrong in a very short time. Several rockets ignited late, a fuel system failed, and the main engine shut down sooner than planned. Fortunately, however, the R-7 was a massive rocket capable of hoisting into orbit a payload much bigger than the diminutive satellite. Despite the glitches, it continued on course. When the upper stage of the rocket reached an altitude of 142 miles, the engine shut down. At that point, the rocket was traveling away from Earth at precisely the same speed that it was being pulled back, which, translated, meant that it was in orbit. The protective cone was jettisoned, and a beautiful shiny ball separated from the rocket. Four antennas automatically sprung into place and two radio transmitters began to beam a meaningless but incredibly meaningful message back to Earth.

They called it Sputnik Zemlyi, which means traveling companion of the world. Its second name was quickly dropped and everyone would forever know it as Sputnik. For the rocket generation, it would be the defining moment of their lives.

The news broke at a party at the Soviet Embassy in Washington. Despite the fact that Cold War distrust seemed to pervade almost every encounter between Russians and Americans, the reaction was almost universally one of excitement and awe. "I've just been informed . . . that a Russian satellite is in orbit," Berkner told the assembled guests. "I wish to congratulate our Soviet colleagues on their achievement." Someone found a shortwave radio and tuned it to Sputnik's beep. The crowd listened in stunned silence. Joseph Kaplan, chairman of the U.S. delegation for the IGY, called the news "fantastic."[19] Everyone assumed that politics could actually be kept out of space.

In one dramatic act, the Soviet Union had defined the terms of modernity. They had surged ahead in space because of their success in marshaling technological change for purposes of the state. During the 1930s, when Western governments had cut back on state-sponsored scientific research, the USSR steadily increased its research and development (R&D) spending. By the 1950s, the American budget for military research, expressed as a percentage of gross national product, was a fraction of that of the Soviet Union. The United States would respond to Sputnik by seeking to imitate the USSR. Research and development would increasingly be managed for purposes of the state. Funding would increase massively. This obsessive quest to imitate was inspired not because Sputnik was important, but rather because it seemed so. Americans decided that the best way to beat the Russians was to become more like them.

Eisenhower had desperately tried to avoid this scenario. For him rockets were simply tools. They might be useful for protecting America, but they should never become ends in themselves. Symbols might look impressive, but they provided a thin shield. Eisenhower feared that rampant science, allied to an alarmist military, would ruin what was good about America. Just after the shock of Sputnik, he told the American people:

> There is much more to science than its function in strengthening our defense, and much more to our defense than the part played by science. The peaceful contributions of science—to healing, to enriching life, to freeing the spirit—these are [its] most important products. . . . And the spiritual powers of a nation—its underlying religious faith, its self-reliance, its capacity for intelligent sacrifice—these are the most important stones in any defense structure.[20]

The president believed that America was a great country which did not need to prove its greatness by shooting chrome balls into the sky. He feared that what he called the "hysterical approach" would bankrupt

America. That, for him, was the tragic dilemma of the Cold War—how to keep dangers at bay without destroying good at home. It was not a dilemma he would ultimately solve.

On October 4, 1957, ten-year-old Stephen King was watching a science fiction movie about outer space invaders at a cinema in Stratford, Connecticut. Something went wrong with the projector, causing the audience to descend into mayhem. The cinema manager got on the stage, apparently to read the riot act. Instead, looking pale and shaken, he said: "I want to tell you something. I want to tell you that the Russians have put a space satellite into orbit around the Earth. They call it Sputnik."

There was silence for a few moments and then came a voice from the crowd: "Oh, go show the movie, you liar!"[21]

Imagination had given way to reality. A real space age had begun.

5

The Red Rocket's Glare

John Williams of Melbourne, Florida, was having trouble with his garage door. The damn thing was opening by itself at all hours of the day and night. The first time it happened, at 2:00 A.M., he woke up and thought someone was trying to steal his car. He grabbed his gun, ran outside, but found no one. He assumed the thief had fled.

Then it happened again. And again.

It finally dawned on him that something must be wrong with the electronic mechanism that opened the door. He called the firm that installed it and explained the problem. They said not to fret. It was happening to loads of people lately. It was all due to that Russian thing in the sky, you know, Sputnik. That satellite was broadcasting on the same frequency as some remote control garage door openers. Williams put down the phone and pondered one of the weirdest conversations he'd had in his life. He couldn't figure out whether to be relieved—or scared.

Sputnik arrived at a time when the news was already consistently gloomy. Grisly murders dominated the nightly news. A flu epidemic in 1957 would eventually kill 70,000 across the United States. In Little Rock, Arkansas, tensions were high due to the enforcement of new integration legislation at Central High School. Americans were already feeling despondent, even before the arrival of that Russian thing in the sky.

At the New York public library, staff noticed an unusual demand for articles on rocketry and for the works of Jules Verne.[1] The American Rocket Society claimed that, because of Sputnik, the ranks of amateur rocketeers had swelled to more than 10,000. At hospitals around the country, emergency room doctors encountered an astonishing number of young boys who had been injured while playing with their potentially lethal toys.* The problem got so bad that the state of New Jersey attempted to ban the hobby.[2]

---

* The society reported 162 injuries during the period June to December 1958 (NASA History Office, Folder 6718).

The very first opinion polls on Sputnik showed a surprisingly calm reaction. Some people were frightened but most were not. Some felt that it was a blow to U.S. prestige, but more did not. Gallup found that 61 percent of those polled were certain that the next great leap forward in space would be American.[3] In an ideal world, responsible journalists might at this point have sought to educate the public on what the satellite actually meant. But that didn't happen. Instead of analyzing the news, the press created it. On October 6, the *New York Times* quoted an unnamed "naval scientist" who claimed that "the very fact that the satellite is whizzing around the Earth would indicate that the Russians are ahead in rocketry. . . . It means that the Russians must have the intercontinental ballistic missile."[4] On the following day, the same paper asked: "Is the world faced with a radical change in the military balance of power . . . presumably to be measured in months or a small number of years, when the Soviet Union has enough such missiles to place every major United States city and base under threat of annihilation?"[5]

Life told its readers that it was time to panic. "Let us not pretend that Sputnik is anything but a defeat for the United States," the magazine claimed on October 21.[6] Thanks in part to Hollywood, space was widely seen as a threatening place from which danger came. Green creatures with more arms than legs wielded ray guns and spat poisonous slime. Aliens never smiled. In space, gravity didn't operate; speeds were astronomical; rules (of physics and law) did not seem to apply. Anyone who could master space was obviously superior to anyone stuck on Earth. The reaction to Sputnik was conditioned by these assumptions. A hysterical public, egged on by an ignorant and irresponsible media, engaged in an orgy of fear.

Exacerbating the problem was the fact that the CIA and the National Security Council (NSC), both of which should have known better, advised the White House that the time had come to panic. A chorus went up: the United States had to respond *now*. "If the newspapers printed a dispatch that the Soviet Union planned sending the first man to hell," a disgusted Admiral Hyman Rickover remarked, "our federal agencies would appear the next day, crying, 'We can't let them beat us to it.'"[7] *Pravda* took delight in noticing how U.S. senators were "showing signs of hysteria."[8] Eisenhower tried to calm American nerves but could not compete with the peddlers of panic. "[Sputnik] does not rouse my apprehensions, not one iota," he claimed, "They have put one small ball into the air."[9] Unfortunately, in trying to appear calm, the president managed only to seem old-fashioned.

Arthur C. Clarke called Sputnik "one of the greatest scientific advances in world history." In fact, the science behind Sputnik was pretty basic. The satellite, which weighed 184 pounds, was a spherical object about the size of a large beach ball with four antennas protruding from it. Thrown into orbit by a very large rocket, it took around 96 minutes to go around the Earth. Its distinctive feature was not its size, shape, or weight, but its sound. In the key of A-flat, it sent a continuous beep-beep-beep back to Earth. The beeps were as long as the pauses in between, about three-tenths of a second. "Listen now for the sound that forevermore separates the old from the new," the announcer on NBC radio said with unabashed hyperbole.[10] The simplicity of the sound was itself cause for panic, since many assumed that it could not possibly be as simple as it seemed. A deep, dark message was probably being broadcast, if only the code could be broken. Experts from the CIA, the Defense Intelligence Agency, and all the branches of the military sacrificed sleep in order to decipher the code. All that expertise and hard work yielded one conclusion: Sputnik was saying beep, beep, beep.

Many assumed that the satellite had to have a camera on board. It seemed inconceivable that the Russians would put aloft an object capable of nothing more than transmitting an innocent, repetitive signal back to Earth. When an IGY delegate in Washington formally assured Americans that Sputnik had no eyes, and that there was no secret code, some were satisfied, while others felt their suspicions deepen. Charles Wilson, the outgoing secretary of defense, tried desperately to calm nerves. Sputnik, he said, was "a neat scientific trick . . . nobody is going to drop anything down on you while you're asleep, so don't worry about it."[11]

Sputnik was quickly incorporated into the American vocabulary, as a term of wonder, of derision, and of surprise. A grocer marketed Sputnik oranges which were, quite naturally, "out of this world." To "go sputnik" was an early way of saying to "go into orbit," or to act completely daft. One political wag remarked that Sputnik was Russian for "no tax cut for the Americans next year." Golfers inevitably adopted the name to describe an extremely high shot, particularly if unintended, such as off the tee.[12]

When Americans get worried, their fear quickly spreads to the stock market. On October 7, the Dow Jones index declined 6.32 points and two weeks later the market experienced its largest one-day loss in two years. At the same time, "rocket shares" surged. The aeronautics industry, still hurting from Eisenhower's efforts to cut military spending, sensed brighter days around the corner. Planned layoffs were abruptly canceled.

Imaginative entrepreneurs saw opportunities. Almost immediately, one confectioner came out with a Sputnik lollipop. Restaurants served Sputnik doughnuts and Sputnik burgers—the latter with an orbiting pickle suspended on a toothpick. Just in time for Christmas, a toy manufacturer rushed into production a "pednik," a space vehicle powered by a foot pedal. One clever designer came up with a spherical container that held ice cream. One antenna went into a Coke bottle and the other served as a straw. Presto, a Sputnik ice cream float. There was also a Sputnik lamp and a fly killer shaped like the satellite, not to mention the inevitable Sputnik dance and hairstyle, both of which were otherworldly.

To most Americans, the idea that the Soviets would be the first to put a satellite in orbit seemed inconceivable. As the Russia-watcher John Gunther later reflected, "for a generation it had been part of the American folklore that the Russians were hardly capable of operating a tractor."[13] Assumptions concerning Soviet technological backwardness were so widely believed that some Americans desperately sought alternative explanations for the launch. One rumor held that the satellite was built by Peter Kapitza, the gifted atomic scientist who had spent many years at Cambridge University before the war. The story was taken up by the *American Mercury,* which published an article titled "Britons Helped Win the Satellite Race."[14]

In a similar vein, the *Chicago Tribune* claimed that the technology was 100 percent American, and that it had been part of the package of secrets passed by Julius and Ethel Rosenberg to the Russians in 1947. The paper found a former prosecutor on the Rosenberg case named Myles Lane who confirmed that plans for satellites were among the contraband documents. On the strength of that story, the Senate Internal Security Committee decided to launch an investigation. They went to Lewisburg Penitentiary in Pennsylvania to interview the former spy David Greenglass, but he had little to add to the mystery. It finally transpired that the only thing that could have been passed to the Russians was the 1946 RAND report, "Preliminary Design of an Experimental

World-Circling Spaceship," which would not have told Korolev anything he didn't already know. What no one asked was, if Sputnik was essentially American, why had the United States so far failed to get a satellite in orbit?

Out in Huntsville, Wernher von Braun and Major General John Medaris saw the launch of Sputnik not as a giant leap for mankind, but as a massive humiliation for America. The news broke when Neil McElroy, the newly designated secretary of defense, was touring the Huntsville facilities.* Gordon Harris, the public affairs officer at the base, burst into the room and made a beeline for Medaris, who was enjoying a predinner drink with McElroy and von Braun. "General," Harris interrupted, "it has just been announced over the radio that the Russians have put up a successful satellite. It's broadcasting on a common frequency, and at least one of our local 'hams' has been listening to it."[15] According to Medaris, von Braun reacted as if he "had suddenly been vaccinated with a Victrola needle." He pleaded with McElroy to be given the freedom to fire his rockets. "We knew they were going to do it!" he shouted. "Vanguard will never make it. We have the hardware on the shelf. For God's sake, turn us loose and let us do something. We can put up a satellite in sixty days, Mr. McElroy! Just give us the green light and sixty days!"[16] During the dinner, McElroy was seated between von Braun and Medaris, who, like tag team wrestlers, hit him with a barrage of ABMA propaganda. The tirade was occasionally interrupted by further news of Sputnik, which added exclamation points to their arguments. In the morning, McElroy went back to Washington with von Braun's message drilled into his brain.**

Eisenhower felt that he had more urgent matters on his plate than orbiting Russian gadgets. When the news broke, he was at his farm in Gettysburg, Pennsylvania, discussing the crisis in Little Rock. Since Governor Orville Faubus had refused to implement a federal ruling to desegregate the state's schools, Eisenhower was preparing to order the Arkansas National Guard to intervene. The scenes from Little Rock stood in stark contrast to the news from the sky. The Russians had Sputnik, the ultimate symbol of modernity, and the Americans had Faubus,

* McElroy would not assume office until 9 October.
** Medaris had told McElroy that, to be on the safe side, the army should perhaps be given ninety days, not sixty.

the icon of bigotry. Keen to drive the point home, Radio Moscow announced the precise moment when Sputnik passed over Little Rock.*

The president steadfastly refused to panic. "I see nothing at this moment . . . that is significant . . . as far as security is concerned," he announced at a press conference on October 9.[17] He played five rounds of golf during the week after the launch, perhaps to drive home the suggestion of calm. Unimpressed, the *Birmingham News* headlined one of its Sputnik articles: "Ike Plays Golf, Hears the News." Mennen Williams, the governor of Michigan who fancied himself a poet, merged golf and Sputnik into two quatrains:

> *Oh little Sputnik, flying high*
> *With made in Moscow beep,*
> *You tell the world it's a Commie sky*
> *And Uncle Sam's asleep.*
>
> *You say on fairways and on rough*
> *The Kremlin knows it all,*
> *We hope our golfer knows enough*
> *To get us on the ball.*[18]

Meanwhile, James Hagerty, the White House press secretary, told journalists that the satellite, while of great scientific interest "did not come as any surprise; we have never thought of our program as in a race with the Soviets."[19]

Nor, apparently, did Nikita Khrushchev. The Soviet premier was initially not all that excited about Sergei Korolev's achievement, which merely confirmed what he already knew, namely that his country was rather good at building big rockets. He did not immediately see that something of huge political significance had occurred. His reaction was mirrored in *Pravda,* which on the day of the launch contained only a small reference on the front page and failed to mention the satellite by name. "Artificial Earth satellites," the article droned, "will pave the way to interplanetary travel, and it seems that the present generation will witness how the freed and conscious labor of the people of the socialist society turns into reality the most bold dreams of mankind."[20] After wit-

---

* Governor Theodore McKeldin of Maryland called Faubus "that sputtering Sputnik from the Ozarks" (Dickson, p. 250).

nessing the reaction in the United States, however, the paper devoted banner headlines and nearly the entire front page to Sputnik two days after the launch.

Khrushchev also began to appreciate what had been achieved. His delight increased when he found that, in nations around the world, Soviet stock had risen, while that of the United States had declined. People seemed to take delight in the fact that America had been pushed from its pinnacle. Indian prime minister Jawaharlal Nehru called the satellite "a great scientific advancement." Though that was hardly an original statement, what mattered was not what Nehru actually said, but that he was talking about something Russian. Cairo Radio was more specific about the conclusions to be drawn. The satellite, it was argued, would "make countries think twice before tying themselves to the imperialist policy led by the United States."[21]

India and Egypt were not allies of the United States and therefore could be expected to interpret the event in a decidedly anti-American way. More worrying was the effect Russian gains had upon America's friends, who feared that their protector seemed unsteady on his feet. The U.S. Information Agency discovered that, among America's allies, her prestige relative to the Soviet Union had declined markedly during the Eisenhower presidency. The general opinion was that it would take the United States at least a decade to catch up with the USSR in space. The Republicans, fearful of the political impact of the USIA report, buried it.*

Most Americans were desperately ignorant about outer space and about what their country was up to out there. One opinion poll revealed that before Sputnik, 54 percent of the American people had never heard of a satellite.[22] "We didn't know what Sputnik was," Joe McRoberts, a civil defense administrator in Michigan recalled. "All the attention at the time had been on missiles, and we seemed to be doing very well there. And then, all of a sudden, here's this thing going round and round the Earth and not coming back down. All of your life you've been taught that what goes up must come down."[23] Even in Maryland, where the Glenn L. Martin Company was building the Vanguard, only 15 percent of the population showed an awareness of the program.

Ignorance reinforced panic. A month after the launch, opinion polls showed that a majority of Americans considered Sputnik a blow to their

---

* It nevertheless managed to leak out just ten days before the 1960 election.

nation's prestige and a roughly equal percentage believed that the United States was behind the Soviets in space research, a significant number believing the gap "dangerously" large. These feelings were validated by G. Harry Stine, whose book *Earth Satellites and the Race for Space Superiority* (1957) had come out before Sputnik. Interviewed by a Denver newspaper, Stine, an employee of Martin, remarked: "We have known in the rocket business for a long time that the Russians were pretty sharp. . . . We lost five years between 1945 and 1950 because nobody would listen to the rocket men. We have got to catch up those five years fast or we are dead."[24] That article appeared on the Saturday after Sputnik went aloft. On Monday, Stine was fired.

Professor A. A. Blagonravov, the chief Soviet scientist at the IGY, assured Americans that "there is no danger to any of the peoples of this world from this Earth-circling man-made Moon."[25] But who was going to believe a Russian? Many Americans concluded that their lead in the arms race—and therefore their security as a nation—had been erased in one dramatic instant. Khrushchev delighted in stoking these fears. In the wake of the launch, he referred derisively to the doubts that had been expressed when the USSR had earlier announced its possession of ICBMs: "Some people said it was only psychological warfare—that we were only trying to impress people. Now we not only have a rocket that can fly to other countries but we have a satellite that flies around the world. I don't have to point to this with my finger—any idiot can see it . . . they might as well put bombers and fighters in the museum."[26] *Newsweek* felt that the Russians "had already given the word 'satellite' the implications of ruthless servitude." The magazine went on to ask, rather ominously: "Could the crushers of Hungary be trusted with this new kind of satellite, whose implications no man could measure?"[27]

Recalling an earlier time in 1812 when America had come under rocket attack, one clever headline writer referred to the "Red Rocket's Glare." "If the Russians can deliver a 184-pound 'Moon' into a predetermined pattern 560 miles out in space," the *Chicago Daily News* surmised, "the day is not far distant when they could deliver a death-dealing warhead onto a pre-determined target almost anywhere on the Earth's surface." One of the most worrying aspects of this crisis was the way the "authorities"—those supposedly in the know—were driving the panic bandwagon. "I would not be surprised if the Russians reached the Moon within a week," John Rinehart of the Smithsonian Astrophysical Observatory remarked. Edward Teller, father of the hydrogen bomb and America's most outspoken scientist, claimed that the Soviets had won "a battle more important and greater than Pearl Harbor."[28] That analogy would be used repeatedly by journalists and politicians desperate to get a grip on the true measure of the catastrophe.

Soviet scientist Kirill Stanyukovich did tell British reporters that Sputnik was "the necessary first stage in the conquering of interplanetary space. It is the necessary first stage in the flight to the Moon."[29] But that hardly meant that the Russians would be Moon-bound the following week. The idea that they would follow Sputnik with a quick break for the Moon had little logic, but fear is seldom logical. Even less rational was the assumption that occupation of space would give the Russians an enormous strategic advantage. Lyndon Johnson, the Senate majority leader, warned his fellow Americans that, unless they woke up to the problem, "soon, [the Russians] will be dropping bombs on us from space like kids dropping rocks onto cars from freeway overpasses."[30] The idea that bombs could be "dropped" from a space platform was a favorite theme of the alarmists. They seemed to ignore the fact that, if the satellite was orbiting, so too would the bomb. In fact, directing a missile from a moving platform in space to a specific terrestrial point was immensely more difficult than launching a similar weapon from Earth.

Frontier myths were carelessly applied to the space age. "What is at stake is nothing less than our survival," Senator Mike Mansfield warned.[31] The Moon was up there—a commanding vantage point for any ruthless oppressor. As every Hollywood western showed, it was dangerous to let the enemy occupy the high ground, as Johnson repeatedly reminded his listeners. He spoke endlessly about the battle between good and evil to be won in space. His Texan accent gave the message added weight. "Control of space means control of the world, far more certainly, far more totally than any control that has ever or could ever be achieved by weapons, or troops of occupation," he argued. "Whoever gains that ultimate position gains control, total control, over the Earth, for the purposes of tyranny or for the service of freedom."[32]

Sensing an electoral opportunity, Johnson grabbed the crisis by the scruff of the neck. "The issue is one which, if properly handled, would blast the Republicans out of the water, unify the Democratic Party, and elect you President," an aide told the ambitious senator. Before Sputnik, the Democrats were mired in gloom. The segregation issue, which had split the party, seemed likely to destroy their chances of regaining the

White House. Then, suddenly, here came a crisis to distract attention from racial problems and bring all elements of the party together. Johnson told Democrats that they had "an incomparable opportunity to save the nation and the world."[33] "LBJ was eager to get out in front in space because it was the new national toy," a Republican recalled. "He was trying . . . to become President of the United States. . . . You do like Robespierre—'There goes the crowd. I must get in front of them, I'm their leader.'"[34]

"The Roman Empire," Johnson claimed, "controlled the world because it could build roads. Later—when it moved to sea—the British Empire was dominant because it had ships. In the air age we were powerful because we had airplanes. Now the communists have established a foothold in outer space."[35] The history lesson might have been crude and simplistic, but the American people lapped it up. On another occasion, he claimed: "From space the masters of infinity would have the power to control the Earth's weather, to cause drought and flood, to change the tides and raise the levels of the sea, to divert the Gulf Stream and change temperate climates to frigid."[36] He promised that, whatever the administration's space budget, he would convince Congress to increase it, a promise that alarmed the fiscally conservative Eisenhower.

The good populist knows the importance of slogans—witness William Jennings Bryan with his "cross of gold." Johnson set his staffers to work thinking up a clutch of catchphrases designed to inflame the passions of the people. These slogans became a substitute for policy. The public, encouraged to see politics in simple terms, demanded simple solutions, where none existed. The mass media cooperated in this lemming-like march. Johnson had only to warn that the United States was in a race and before long the American people started running. The direction hardly mattered.

Sputnik had made only a few orbits of the Earth before Johnson was on the phone to his Senate colleagues demanding hearings on America's role in space, and the related issue, the supposed "missile gap." The Inquiry into Satellite and Missile Programs began on November 25. It was supposed to be nonpartisan, but witnesses and testimony were carefully arranged to pour scorn on the government and to showcase the statesmanlike Johnson. One clever trick was to comb classified government reports detailing existing satellite or rocket programs. The committee would then recommend these already functioning programs to the government and then take credit for the idea when the lid was eventually lifted. An ignorant public naturally concluded that the government was acting on the advice of Johnson.

One of the "expert" witnesses was Teller, who, five years earlier, had helped convince the United States to build a thermonuclear bomb for which there was no real purpose. "Shall I tell you why I want to go to the Moon?" he asked the assembled senators.

> I don't really know. I am just curious. . . . If you asked me about ballistic missiles in 1945 or 1946, I would have said, "Let's do it and let's do it fast," and then you would have said, "In what particular way will you apply this in a possible war?" and I would have told you, "I don't know, but once we make it we will find some use." And I think going to the Moon is in the same category. . . . It will have both amusing and amazing and practical and military consequences. This is how it always was in the world.[37]

J. Robert Oppenheimer once remarked that a scientific discovery is made not because it is useful but because it is possible. Teller was quite happy to transform that principle into government policy. So, too, was Johnson. America would go into space simply because it could.

The Senate hearings drew to a close on January 23, 1958. Over a period of two months, a parade of scientists, engineers, and military figures heaped a torrent of abuse on the Eisenhower administration for its neglect of the nation's security. Looking every bit the statesman, Johnson ended by reminding Americans that the spectacle was all for the good of America, that politics was immaterial when crisis threatened. "I think that all of us remember the day after Pearl Harbor," he said with appropriate solemnity. "There were no internationalists and no isolationists; no Republicans and no Democrats."[38]

In the House of Representatives, Speaker John McCormack warned that the Americans faced "national extinction" if they failed to respond—and respond quickly—to Sputnik. "It cannot be overemphasized that the survival of the free world—indeed, all of the world, is caught up in the stakes."[39] Alarmists reiterated that Soviet achievements would undermine the appeal of liberal democracy among underdeveloped nations. This interpretation was certainly encouraged by Khruschev, who boasted that within fifteen years the Soviet Union would surpass the United States in per capita economic output. As if to confirm that prediction, Johnson warned that "failure to master space means being second best in every aspect, in the crucial arena of our Cold War world. In the eyes of the world first in space means first, period; second in space is second in everything."[40]

Shortly after the Sputnik launch, Senator John Kennedy was in Boston's Loch Ober Café, where the bartender introduced him to Charles "Doc" Draper, a professor at MIT who specialized in rocket guidance. According to Draper, Kennedy insisted that research into rocketry was a waste of money, especially when it came to the question of exploring space. Draper admitted that Kennedy might simply have been playing the devil's advocate, but it was clear that he had no great enthusiasm for the subject.

Kennedy was, however, an ambitious man who had no intention of letting Johnson ride a rocket to the White House. He therefore decided to challenge LBJ for the role as prophet of doom. Echoing the Texas senator, he argued: "Control of space will be decided in the next decade. If the Soviets control space they can control the Earth, as in past centuries the nation that controlled the seas dominated the continents. . . . We cannot run second in this vital race. To insure peace and freedom, we must be first."[41] From his private letters and recorded conversations, not to mention the recollection of his close colleagues, it is clear that Kennedy never remotely believed that Soviet space spectaculars endangered the safety of the United States, but he did sense that there were votes to be won from panic. In 1958, he warned:

> The people of the world respect achievement. For most of the twentieth century they admired American science and American education, which was second to none. But now they are not at all certain about which way the future lies. . . . If the Soviet Union was first in outer space, that is the most serious defeat the United States has suffered in many, many years. Because we failed to recognize the impact that being first in outer space would have, the impression began to move around the world that the Soviet Union was on the march, that it had definite goals, that it knew how to accomplish them, that it was moving and we were standing still. That is what we have to overcome, that psychological feeling in the world that the United States has reached maturity, that maybe our high noon has passed and that now we are going into the long, slow afternoon.[42]

Kennedy sensed that Sputnik had exposed a Republican Achilles Heel. The satellite would allow him to take on the role of Winston Churchill, automatically casting Eisenhower as Neville Chamberlain. The Moon, like China and Korea, could be presented as a domino that must not fall. The Republicans' hold on the White House, which had seemed rock solid, suddenly looked vulnerable. Kennedy made Sputnik into a moral lesson on the perils of Republican complacency. The postwar economic boom had encouraged dangerous self-satisfaction. The Soviets, it seemed, were lean and hungry, the Americans fat and sluggish. Kennedy saw space as a chance to take American politics into new worlds, where achievement would not be constrained by lack of ambition, red tape, or fiscal prudence.

Before long, thanks in part to Kennedy and Johnson, Sputnik became a symbol of American complacency. Renewal of the American spirit would be achieved in outer space. The Boy Scouts of America took up the theme of rebirth, apparently forgetting that their organization is an international one intended to promote brotherhood between boys of all nations. "The best thing that has happened to the United States in our generation is Russia's Sputnik," a publication proclaimed. "It is proving the most effective therapy conceivable for our indifference and negligence, our concentration on comforts and convenience, our devotion to all but our patriotic obligations in this best of all lands."[43]

Leave It to Beaver, that iconic portrayal of white middle class suburban contentment, premiered on CBS the same night as Sputnik, proof that history has a talent for irony. The coincidence was of course accidental, but no less striking. In the new age of space, a high standard of living suddenly seemed an indicator of cultural inferiority. Clare Booth Luce, politician, playwright, journalist, and diplomat, called Sputnik's beep "an inter-continental outer-space raspberry to a decade of American pretensions that the American way of life was a gilt-edged guarantee of our national superiority."[44] Senator Styles Bridges of New Hampshire urged his fellow Americans to "be less concerned with the depth of the pile on the new broadloom rug or the height of the tailfin on the new car and to be more prepared to shed blood, sweat and tears if this country and the free world are to survive."[45] Echoing that sentiment, the outspoken statesman Bernard Baruch remarked: "If America ever crashes, it will be in a two-tone convertible. While we devote our industrial and technological power to producing new model automobiles, the Soviet Union is conquering space. America is worried. It should be."[46] Pouring salt on the wounds, Leonid Sedov, chairman of the Committee of Interplanetary Communications of the Soviet Academy of Sciences, remarked: "America is very beautiful, and very impressive. The living standard is remarkably high. But it is very obvious that the average American cares only for his car, his home and his refrigerator. He has no sense at all for his nation. In fact, there exists no nation for him."[47]

The arguments about complacency had some foundation. Between 1950 and 1958 the Soviet system enjoyed an average annual growth rate in GNP of 7.1 percent. In contrast, the United States, during the same period, had trouble keeping its growth rate above 3 percent.[*] Almost everyone seemed to have an opinion on America's decline, and no one hesitated to voice a view. "I think it is going to take even more emphasis on the need for sacrifice, belt-tightening and renewed dedication, if the American way of life is to be preserved," Billy Graham wrote Eisenhower. "We Americans are growing soft. Our amusements and greed for money is [sic] acting as a sedative. We almost need laws for compulsory scientific training in all our schools, as well as compulsory physical training for all our young people. In other words, we need to touchen [sic] up!" He also felt that "the American people need to be encouraged to look to God who is the source of all our spiritual and moral strength."[48]

Some people began to wonder if perhaps the Soviet system, even though it failed to provide televisions for all, actually worked. Tied to the economic argument was the widely trumpeted claim that the Soviet system was built on fair distribution and equal rights for all, an argument that hit home at a time when rural poverty plagued America and people were fighting in the streets to prevent black children from going to school.[**] The Arkansas senator J. William Fulbright, however, rejected this line of reasoning. He insisted that Sputnik was nothing but an elaborate stunt. "It does not feed their people. . . . It does not convert any-

---

[*] Kennedy made much of these figures during the 1960 campaign, but they were highly misleading. The Russians were recovering from the devastation of World War II, therefore any growth was bound to be profound. In any case, as Kennedy would eventually discover, the Soviet statistics were not entirely accurate.

[**] The Soviets would later capitalize on this argument by making much of the Chuvash origins of the cosmonaut Nikolaev, who went into orbit in 1962, and also by hastening the mission of the first female cosmonaut, Valentina Tereshkova.

one to communism. So far as real prestige is concerned, it is nothing."[49] Unfortunately, Fulbright's calm reason was drowned by the tide of panic.

When asked by legislators to explain what was wrong with American science, von Braun called for a massive injection of money and effort to be directed to the teaching of science. Legislators responded with calls for a Manhattan Project for space and a West Point for science. The Advanced Research Projects Agency was created to make sure that Americans were kept busy with technological projects—not exclusively related to space—designed to make sure that the United States would never again fall victim to an embarrassment like Sputnik. The budget for the National Science Foundation rose from under $50 million in 1958, to $136 million in 1959 and by 1962 had exceeded $250 million. The National Defense Education Act (NDEA) was rocketed through Congress and signed into law by Eisenhower on September 2, 1958. A four-year plan for boosting American education, it provided millions for the purchase of scientific equipment for schools, in addition to loans and grants for those inclined to go into teaching. Before long, the NDEA was giving out scholarships to almost any high school graduate who could present a credible case for wanting to study science at a university.*

All this alarmed the *Wall Street Journal,* which feared an assault upon free enterprise carried out in the name of national security. "The idea that the Federal government can do all things better than local or private initiative is not borne out by the facts," the paper argued. "So who are the defeatists? Those who say that only in freedom can our people show their will, energy, resourcefulness and capacity? Or those with so little faith in freedom that they invite us to start surrendering our liberties in panic?"[50] Eisenhower was trying to make precisely that argument, but his voice was drowned out by that of von Braun, who was looking and acting more like a prophet with every passing day.

Teachers, a long demoralized group, delighted in the attention finally paid to their profession. "Any nation that pays its teachers an annual average of $4,200 cannot expect to be first in putting an Earth satellite into space," one campaigner remarked, apparently unaware that Russian teachers were paid peanuts. Schools deployed a new strategy to get students to do their homework. After Sputnik went up, all the talk

---

* One of the chosen was Theodore Kaczynski, a.k.a. the Unabomber, who would eventually put his skills to a use that the NDEA never quite foresaw.

was about the "fact" that Russian kids went to school for six hours a day, six days a week, and got shorter summer holidays than the Americans. An official report by the Department of Health, Education and Welfare made the preposterous claim that *all* Russian children took five years of physics, four of chemistry, and five of mathematics. Some teachers then embellished these statistics by telling their pupils that the little communists also got four hours of homework each day. As the Hollywood producer David Obst later remarked, "[My] classmates and I became staunchly anticommunist."[51]

The *Washington Post* compared the situation to the state of unpreparedness in 1941. As had happened after Pearl Harbor, "to overcome the scientific and psychological disadvantage, it will be necessary for the Administration first to assert some aggressive leadership and then to provide whatever talent and money are necessary to right the balance." In response, a desperate Eisenhower insisted that his approach to the problem was similar to that which he had implemented to defeat the Nazis. During the war, "it was a good plan, a long-range plan that had been carefully worked out. We went on and won."[52] The point he tried to make, but could never quite get across to the American people, was that the United States "could make no more tragic mistake than merely to concentrate on military strength."[53] Lyndon Johnson was cynically dismissive of these attempts to calm the people's nerves. "I guess for the first time I've started to realize that this country of mine might not be ahead in everything," he remarked. "It is not very reassuring to be told that next year we'll put an even better satellite in orbit, maybe with chrome trim and automatic windshield wipers."[54]

One of the reasons Eisenhower was not inclined to panic was because he genuinely felt that his objectives had been achieved. Quarles, now deputy secretary of defense, told a cabinet meeting that the "Russians have done us a good turn, unintentionally, in establishing the concept of freedom of international space—this seems to be generally accepted as orbital space, in which the missile is making an inoffensive passage."[55] Unfortunately, that explanation was never likely to go down well with the American public.

Under pressure from all sides, Eisenhower tried diligently to defend himself. On October 9, he told a press conference that he saw "nothing . . . that is significant in that development as far as security is concerned, except . . . it does definitely prove the possession of the Russian scientists of a very powerful thrust in their rocketry. . . . The mere fact that this thing orbits involves no new discovery of science . . . so in itself it imposes no additional threat to the United States."[56] He went on to allege that the reason the Russians had been first in space was because they had "captured all the German scientists" in Peenemünde.[57] Secretary of State John Foster Dulles appears to have been the origin of this claim. In a memo to Hagerty, he wrote: "The Germans had made a major advance in this field and the results of their effort were largely taken over by the Russians when they took over the German assets, human and material, at Peenemünde.[58] Needless to say, that raised a few hackles in Huntsville.*

A short time later, a syndicated reporter filed a story to the effect that an unnamed but nationally well known missile engineer in Huntsville had disclosed that his team had offered to help the Vanguard program, but that his offer had been rejected. He went on to say that he could get a satellite in orbit in three to six months. There was no doubt as to the identity of the source. Eisenhower was furious since the disclosure violated a gag order on the subject. Von Braun, the ink hardly dry on his citizenship papers, was feeling confident enough to take on the president.

For some reason, no one bothered to ask the experts about Sputnik, or, rather, they asked the wrong experts—those (like von Braun) who had much to gain from fear. Out at Edwards Air Force Base in the California desert, the air force was testing its X-series of rocket planes. Nearly ten years before Sputnik, on October 14, 1947, Chuck Yeager had broken the sound barrier in the X-1. Yet that accomplishment had hardly raised a stir.** Each new model flew farther, higher, and faster. Yeager wasn't very impressed with Sputnik. He was in Germany serving as a squadron commander on the day the news broke. "I thought to myself, 'What the hell's that thing accomplishing?'" All the Russians had done was put a tin can atop a powerful rocket and then shot the rocket straight up. For the previous decade, Yeager had been learning to fly rockets, with the aim that he, or one of his colleagues, would even-

---

* Von Braun later remarked, rather dryly: "I always considered that the President got some poor advice on that item." When it was pointed out that there were rather a lot of Germans working in Huntsville, one State Department official quipped: "We captured the wrong Germans!" (Von Braun interview, Eisenhower Library; Crouch, p. 92).

** The air force refused to publicize the event until the following June, with the effect that many people ended up believing that the feat had been accomplished by an Englishman, thanks to the 1952 movie *Breaking the Sound Barrier*, starring Ralph Richardson.

tually *fly* one into outer space. The latest in the line, though yet to emerge, was the X-20, or Dyna-Soar (for dynamic soaring). Sputnik, in comparison, looked primitive, proof not of the Russians' technical lead, but rather of their backwardness. Thirty years later, Yeager reflected: "It was our country's response to the event that was important to history, not the event itself."[59]

In January 1958, not long after Sputnik fell from its orbit, Gabriel Heatter of the Mutual Broadcasting System delivered a radio editorial he called "Thank You, Mr Sputnik." It summed up everything Americans had been feeling over the past three months of agony and embarrassment.

> You will never know how big a noise you made. You gave us a shock which hit many people as hard as Pearl Harbor. You hit our pride a frightful blow. You suddenly made us realize that we are not the best in everything. You reminded us of an old-fashioned American word, humility. You woke us up out of a long sleep. You made us realize a nation can talk too much, too long, too hard about money. A nation, like a man, can grow soft and complacent. It can fall behind when it thinks it is Number One in everything. Comrade Sputnik, you taught us more about the Russians in one hour than we had learned in forty years.[60]

Americans had indeed been deluded about the extent of their greatness. But Sputnik, while it might have destroyed that misconception, replaced it with something equally misleading. Delusion gave way to illusion. Many concluded that the cure to America's woes, whatever they might be (and no one was quite sure what they were), lay in outer space. Putting men into tiny tin cans and putting those tin cans on top of gigantic roman candles would, it was thought, allow America to be great again. Comrade Sputnik encouraged a yearning for the freshness and gladness of youth, when problems seemed simple and when fastest and farthest were synonyms for best.

6

Muttnik

Sometime in 1956, an analyst at RAND came up with a really clever way of getting a jump on the Russians. W. W. Kellogg decided that an atom bomb could be attached to a rocket and fired at the Moon. Timed to explode just before hitting the lunar surface, it would send an astounding visual display back to Earth. Within RAND, the idea didn't go very far, since analysts were busy with more important matters and, in any case, the United States at that point couldn't put a grapefruit into orbit, much less send an atom bomb to the Moon. But the idea didn't go away.

Two years later, William Pickering at the Jet Propulsion Laboratory argued that such a stunt would have valuable scientific merit, given that it would "shower the Earth with samples of surface dust." This dust (radioactive, it should be noted) could be studied, yielding valuable conclusions about the origins of the Moon. Pickering also thought that the mission would have "beneficial psychological results," namely scaring the crap out of the Russians by demonstrating the range and accuracy of American missiles. Pickering insists that it was never more than "coffee table talk," yet some very intelligent people spent valuable time exploring the possibility. Unbeknownst to those who had originally proposed the idea, a secret government study was conducted at the Armour Research Foundation in Illinois. The physicist Leonard Reiffel was asked to look into feasibility. From May 1958 to January 1959, Reiffel's team (which included the young Carl Sagan) studied the likely effect of a Hiroshima-sized bomb on the Moon's surface, and the visual display of such an explosion, if seen from Earth. "As these things go, this was small," Reiffel later confessed. "It was less than a year and never got to the point of operational planning. We showed what some of the effects might be. But the real argument we made, and others made behind closed doors, was that there was no point in ruining the pristine environment of the Moon. There were other ways to impress the public that we were not about to be overwhelmed by the Russians."[1]

. . .

One wonders if the idea was ever mentioned to Eisenhower. Probably not, since the president abhorred stunts. His quiet tranquillity had long been considered one of the most reassuring qualities of his presidency. Unfortunately, however, once Sputnik went into orbit, that quality seemed more like complacence. It did not help that the chief of Naval Research, Admiral Rawson Bennett, scorned the satellite as "a hunk of iron almost anybody could launch."[2] An attempt by the White House chief of staff Sherman Adams to dismiss the satellite also backfired badly. "The serving of science, not high score in an outer space basketball game . . . is our country's goal," he said.[3]

A second and even more impressive shot demonstrated that the Soviets were rather good at basketball. On November 3, 1957, they launched Sputnik II, a satellite weighing about 500 kilos, or about the size of a small car (something virtually unknown in America). Even more remarkable than the weight was the fact that the capsule contained a dog named Laika, and the systems necessary to keep it alive for a short time.[*]

Poor Laika had a one-way ticket to space. She was originally given food and oxygen to last just 100 hours, but, due to the loss of a thermal shield at launch, she didn't even manage that long. She stayed alive as the heat in the capsule rose, finally expiring on the fourth day, much to the dismay of dog lovers around the world. For many Americans, the incident demonstrated once again how barbaric the Soviets were, but it was rather painful that these dog despisers were so far ahead in the space race.

American papers dubbed the craft Muttnik or Poochnik. One restaurant began offering the "Sputnikburger with small dog at no extra charge."[4] Humor could not, however, obscure a profound unease. The logic of Russian intentions seemed clear: where dog went, man would follow. Robert Gilruth, at the time working on rocket-powered craft at Langley Laboratories, confessed that, before Sputnik, "I never thought of flying people in space." Then came Laika and "I said to myself and my colleagues, 'This means that the Soviets are going to fly a man in space.'"[5] This was confirmed, though not officially, by the First Secretary at the Soviet Embassy in London, who answered protesters by arguing that the dog was testing the safety of space travel for humans.

* The name Laika carries a certain historical resonance only because few people outside Russia speak Russian. The word actually means "barker"—proof that the Soviets put a great deal more effort into content than style.

What no one asked, however, was what purpose man (or dog) could possibly serve in outer space. Once Laika went into orbit, Soviet and American perceptions of space changed radically. Up until that point, the important issue was the Soviet ability to lift very heavy objects into orbit. That capability seemed to threaten U.S. security, as Johnson delighted in pointing out. Putting a living organism in space did not remotely worsen the threat. It did, however, introduce very expensive complications into the space equation. At the very moment when the heavens opened for the Soviets, they decided, for egotistical purposes, to head down the cul de sac of manned space travel. Where they led, the Americans followed—rather like a blind man being led by a dog.

In the November edition of *Missiles and Rockets,* editor Erik Bergaust wrote "An Open Letter to President Dwight D. Eisenhower." The magazine, standard reading for every rocket enthusiast, took the extraordinary step of announcing the letter beforehand with a full-page ad in the *New York Times.* The editorial expressed all the emotions the president feared. "This is the age of science," Bergaust warned Eisenhower. "This is the era of intellectual, uninhibited thinking. Tomorrow is here. And you, as the leader of the greatest nation on Earth, must see to it that this nation will be out in front as mankind advances into the space age." Bergaust called for a cabinet-level science adviser, a coherent space program, a new space agency, and missions to the Moon, Venus, and Mars.[6]

Eisenhower realized that he could not simply ignore the American people's feelings of inadequacy. On November 7, before a radio and television audience, he delivered the first of his "Science and Security" talks. The speech was delivered while sitting at his desk, with the nose cone from a Jupiter rocket nearby. The specimen was the first ever object recovered from space, but that was hardly enough to appease Americans who wanted space dogs, not charred nose cones. Perhaps, in an oblique reference to Laika, Eisenhower was trying to send the message that Americans, at least, had the ability to recover things from space, instead of abandoning them.

Since the president did not, in truth, have much to reveal, the speech was mostly verbiage. He did mention that steps would be taken to address the inter-service rivalries that had delayed progress in the missile field, but this only added credence to rumors that internal animosities had been to blame for America's poor performance. In response to critics who derided the state of scientific education in the

United States, the president announced that he was appointing James Killian of MIT the first White House science adviser.

On the following day, more momentous news was released. The Department of Defense issued a short press statement announcing that the army would be allowed to proceed with an attempt to launch a satellite. The release stressed that the army rocket would supplement the existing Vanguard project and that "there is every reason to believe Vanguard will meet its schedule to launch later this year a fully instrumented scientific satellite."[7] Eisenhower tried to give the impression that the army was being asked to start on a new satellite program, but those in the loop knew better. The columnist Drew Pearson correctly surmised that the hardware for an army launch had been sitting in a warehouse in Huntsville for some time. Medaris denied the allegations, but not vehemently.

By letting Medaris and von Braun off their leash, Eisenhower had, in a stroke, confirmed that he now considered space a race, and that, because the United States was already seriously behind, his earlier scruples about keeping the military out of the satellite venture no longer applied. The president had surrendered what little control he still retained in the space arena. The Russians, not to mention Medaris and von Braun, were pulling him by the nose into space.

Two months after Sputnik, on December 6, 1957, the United States made its first attempt to place a satellite in orbit. The Vanguard team had a relatively new rocket that had not been fully tested. They had intended that TV-3, the fourth test launch, and the first to involve a fully operational rocket with all three stages ready to fire, would take place in early December. A satellite would be installed on top of the rocket, but the test itself would remain experimental—not, in other words, an official attempt to go orbital. Project director John Hagen felt that no publicity should be given to this test, but if it achieved orbit then it could be appropriately celebrated. The real attempt would take place with TV-5.

Unfortunately, Hagen did not bargain on the pressure emanating from the White House. On October 11, Hagerty announced that an attempt to reach orbit would be made in December, in other words with TV-3. Hagen took this as a presidential directive to put a satellite in orbit before the end of the year. Meanwhile, the press started to gather down in Cocoa Beach. J. Paul Walsh, deputy director of the Vanguard program, tried desperately to bring calm to the frenzied air of expectation.

"We'll be pleased if it goes into orbit," he announced. "We'll not be despondent if it does not."[8]

The launch was originally scheduled to take place on December 4, but bad weather and a variety of technical glitches caused a two-day delay. Thousands of spectators waited impatiently for the precise moment America would enter the space age. High atop the slender rocket was perched the payload, a minuscule 3.2-pound satellite assigned the task of delivering that all-important "beep" from outer space.

At 11:44:55 A.M. on Friday, December 6, 1957, the slender rocket rose slowly from its launch pad. After two seconds, and at an elevation of about four feet, it abandoned the struggle. It shuddered, then collapsed in a fiery heap. Kurt Stehling, who worked on the propulsion system, recorded his impressions:

> It seemed as if the gates of hell had opened up. Brilliant stiletto flames had shot out from the side of the rocket near the engine. The vehicle agonizingly hesitated for a moment, quivered again, and in front of our unbelieving, shocked eyes, began to topple. It sank, like a great flaming sword into its scabbard, down into the blast tube. It toppled slowly, breaking apart, hitting part of the test guard and ground with a tremendous roar that could be felt and heard even behind the two-foot concrete wall of the blockhouse and the six-inch bulletproof glass. For a moment or two there was complete disbelief. I could see it in the faces. I could feel it myself. This just couldn't be.[9]

Playing on the complacent America theme, journalist Tom Wolfe later described how the rocket appeared "to sink very slowly, like a fat old man collapsing into a Barcalounger."[10] As a result of the explosion, the grapefruit-sized satellite broke free and rolled under a nearby shrub, where it began to broadcast its forlorn beep to the world.

"A shot may be heard around the world," remarked the Louisville *Courier-Journal*, "but there are times when a dud is even louder." Headline writers had great fun. The American satellite was christened Stayputnik, Flopnik, Oopsnik, Pfftnik, and Sputternik. The West German press dubbed Vanguard "Spaetnik"—"spaet" meaning "late" in German.[11] At the United Nations, the Soviet delegation formally offered financial aid to the United States as part of a program of technical assistance to backward nations. What no one quite realized was that the rocket had performed pretty much to expectation. The launch was supposed to be a test, not the real thing, but the impatience of the government had transformed it into something more dramatic. The R-7 rocket that carried Sputnik aloft had failed six times before its momentous success on October 4. But few people were interested in details.

Over the next couple of months, the air force tried to sort out Vanguard's problems. At one point near the end of January 1958, the rocket came within fifteen seconds of launching when the mission was aborted. None of this was, however, reported. The first Vanguard failure taught the rocket men a very important lesson about news management. After it, the air force proposed a deal with reporters whereby they would be briefed fully about the good news as long as they kept quiet about the bad. Accept the deal, or find access to our facilities severely restricted, the message went. Thus began a beautiful friendship. Over the next ten years, a relatively small group of space reporters cooperated fully in relaying a well-prepared story to the American public.

Though Laika was long dead, Sputnik II continued to broadcast its signal until early December, thus pouring salt on American wounds. Four months later, on April 14, 1958, observers on the East Coast from Connecticut to the Caribbean witnessed an object streaking across the sky at incredible speed. Those in the North saw several objects separate themselves from it, and these then appeared to fall into formation behind the "mother ship." The spectacle was extraordinary, inspiring awe and panic among those who witnessed it. Many concluded that they were watching an attack fleet of flying saucers. UFOs had been in the news a lot lately, perhaps because so many people were spending so much time looking at the sky.

And then, on the following day, came the explanation. The light show had been provided by Sputnik II and poor Laika, burning up on reentry into the atmosphere.

Meanwhile, behind the scenes, the American military busily dreamed up uses for space, and, specifically, the Moon. In an address to Washington's Aero Club in late January 1958, Brigadier General Homer Boushey, USAF, outlined two possible uses of the Moon: as a missile base, and as an observatory to spy on the Soviet Union. It would also be possible, he surmised, to place the launchers on the dark side of the Moon, thus making them undetectable by the Soviets.

Boushey had in mind an utterly dependable second-strike capability. The Russians, he argued, would henceforth realize that, if they attacked the United States, their action would be observed from the Moon and followed, forty-eight hours later, by "sure and massive retaliation." If the Soviets, in an attempt at first strike, tried to take out the lunar launchers first, they would have to wait two and a half days before then attacking America, more than enough time for the United States to flatten the USSR. He went on to argue that such a defensive stronghold would eventually provide the foundation for a more sophisticated settlement. "With man and his intelligence once established upon the Moon the possibilities of construction and creation of an artificial environment are virtually unlimited." All the natural resources for such a base, he thought, could be extracted from the Moon itself. All of this implied, of course, that the United States could beat the Soviets to the Moon, and then claim it as their sovereign territory. "We cannot afford to come out second in a territorial race of this magnitude," he argued. Then, without a hint of irony, he added: "This outpost, under our control, would be the best possible guarantee that all of space will indeed be preserved for the peaceful purposes of man."[12]

Boushey's reasoning had more holes than the Moon has craters. He never quite explained how a nuclear deterrent placed on the Moon could be any more effective than one on Earth, or indeed whether, if advantages could be derived, they were worth the enormous cost involved. Since most nuclear planners at this time were thinking in terms of a contest with the Soviets lasting hours, not days, missiles arriving from the Moon would probably find that the war was over and the world a smoking ruin. In other words, Moon missiles could never be more than exclamation points to a sentence announcing the end of the world. As Lee DuBridge of the California Institute of Technology commented on hearing Boushey's proposals: "Such plans are nonsense. . . . Why transport a hydrogen warhead 240,000 miles to the Moon just to shoot it 240,000 miles back to Earth when the target is only 5,000 miles away? If you launched a bomb from the Moon, the warhead would take five days to reach the Earth. The war might be over by then."[13]

Boushey, however, was not alone in spouting such lunacy. "I would hate to think that the Russians got to the Moon first," General Dwight Black, the air force director of guided missiles and special weapons told Congress in 1959. "The first nation that does will probably have a tremendous military advantage."[14] Around the same time, the army submitted to ARPA a list of requirements for space transportation and combat which included an interspace vehicle, a space patrol vehicle, a space forward command post, a manned lunar outpost, a lunar assault vehicle, and a lunar surface vehicle.

Civilian contractors were called in to give substance to Buck Rogers–type fantasies of space war. The air force asked Boeing, Douglas Aircraft, General Electric, and a number of other firms to formulate a detailed plan for "a site where future military deterrent forces could be located." Such a site "has potential to increase our deterrent capability by insuring positive retaliation."[15] The proposed permanent lunar base would be built under ground and staffed by twenty people on tours of duty of perhaps seven to nine months. The cost, the contractors forecast, would be no more than $20 billion. Ambition, like self-belief, was limitless. A lunar soil sample, they predicted, could be brought back to Earth in November 1964. The first manned landing would take place in August 1967; establishment of a temporary base three months later. The whole thing would be completed by December 1968 and operational by the following June. Once established, it would be serviced by monthly flights.

The army was even more ambitious. In June 1959, Army Ordnance Missile Command submitted a four-volume report to General Maxwell Taylor, Chief of Staff, outlining the feasibility of a lunar base. All the same flawed reasons for the base were given, all resting on the automatic assumption that a lunar base would make the United States strategically more secure and that greater security was, by definition, always a bargain, regardless of cost. The scheme, the report maintained, should have "an authority and priority similar to the Manhattan Project." Failing to take appropriate action would be "disastrous to our nation's prestige and to our democratic philosophy."[16] The base itself, similar to that outlined in the air force study, would be assembled in space from equipment placed in orbit through regular Saturn rocket launches—147 of them, to be precise. It could be completed by November 1966, after which, during the first full year of operation, an additional sixty-six Saturn launches would be required. Despite the fact that it would require over 210 separate rocket launches, the total cost would not exceed $6 billion.

All these ideas burned up in the atmosphere of Eisenhower's good sense. On the subject of space as a future theater of war, his Science Advisory Committee remarked: "Most of these schemes . . . appear clumsy and ineffective ways of doing a job. . . . Even the most sober proposals do not hold up well under close examination.[17] To the president they seemed confirmation of the danger of allowing soldiers and scientists near each other.* He could not discern any solid strategic justification for a lunar military base, and, in any case, was determined to keep war out of space.

Eisenhower had to watch helplessly as the military-industrial complex took control of American life. The weapons industry was an octopus whose tentacles held politicians, academics, and financiers in a steely grip. Fear bred fear and contracts begat contracts. Universities provided the science, soldiers the rationale, and government the funding for new, ever more expensive, projects. Since science was equated with progress, any new development was automatically assumed to be necessary and beneficial, either for the security of the nation or for the well-being of its citizens. To oppose this progress was seen as dangerous, even unpatriotic. But the more advanced the science, the more difficult it was to judge its worth. As the fuss over Sputnik demonstrated, it was automatically assumed that the definition of a great society was one which produced great science. Utility did not need to be measured. The really clever tactic of American technocrats, and the trait that so annoyed Eisenhower, was that they never precisely defined the objectives of their research. Defining an objective made a project finite, whereas the whole point was to create a climate in which projects, and budgets, could stretch into infinity, each one "improving" on the previous.** The value of R and D was judged by the size of the budget, not by what it produced.

The president was identified with Vanguard's failure and therefore had to share its ignominy. It was a civilian project—intentionally so, since Eisenhower wished for the moment to keep the military out of space. But ABMA shouted to the hilltops that it had a rocket which, if not for the president's ridiculous scruples, would have put an American satellite in orbit in 1956. Von Braun did not hesitate to talk up Juno

* General Bernard Schriever of the Ballistic Missile Division hinted that the idea was never really a serious one, but did have an ulterior purpose: "We had no military arguments. People tried to dream up things we could do on the moon from a national security standpoint and nothing tangible ever came out of it. We felt this would put a focus on the space program; it would accelerate technology. We knew there were a lot of applications in space that did have security implications and we wanted to have a major effort going" (Logsdon, p. 47).

** As will be seen, NASA administrator James Webb hated the idea of a lunar mission precisely because it defined an objective. He wanted Americans to throw their weight behind "the space program," an ideal without limits.

(the civilian version of the Jupiter missile) whenever a reporter entered his orbit. For Eisenhower, behavior of this sort bordered on treason. "When military people begin to talk about this matter and to assert that other missiles could have been used to launch a U.S. satellite sooner," the president remarked, "they tend to make the matter look like a race, which is exactly the wrong impression."[18]

Von Braun got his chance on January 31, 1958. On that day, a 15-kilo Explorer satellite was launched aboard a Juno rocket. Each stage took the rocket higher and faster until it reached an orbital speed of 18,000 miles per hour. A short time later, the antennae of the satellite deployed and it began sending messages back to Earth. Hagerty called Eisenhower just after midnight on February 1 to relay the news. "That's wonderful," the President replied. "I sure feel a lot better now." He hesitated, then added: "Let's not make too big a hullabaloo over this."[19] Neither crises nor triumphs in space suited his purposes. "The United States has successfully placed a scientific Earth satellite in orbit around the Earth," he announced that morning in a statement carefully calculated to downplay the significance. "This is part of our participation in the International Geophysical Year."[20] The army's role was hardly mentioned.

Vice President Richard Nixon told reporters that Explorer was a manifestation of the administration's desire to "develop space exploration in the cause of peace rather than the wastage of war."[21] They were fine words, but rather disingenuous. Juno was essentially an intermediate-range ballistic missile and a direct descendant of the V-2. Its guidance systems had simply been altered so that it went straight up, instead of on a ballistic trajectory. Most Americans, however, did not remotely care that they'd ridden to space on an ersatz German rocket built by a former Nazi. Breathing a deep sigh of relief, they celebrated the fact that the United States had at last started running in the space race. No one bothered to ask about the purpose of it all.

Von Braun felt that Juno's success gave him every right to make a hullabaloo. He realized that an ambitious space program, a hugely expensive venture, required a massive commitment from government and people. The best way to motivate the people (and through them, the government) was through fear. When asked if Sputnik implied that the Russians would be able to hit Washington with a hydrogen bomb, von Braun, without hesitation, replied "yes."[22] In an endlessly repeating refrain he warned that the United States was in mortal danger because of the Russian "lead" in space. Speaking in Chicago on February 17, 1958, he warned: "The Damoclean sword menacing free people consist[s] of a monstrous destructive force inherent in automatic delivery systems, capable of transporting thermonuclear warheads thousands of miles, in *any* weather, across all geographic and political barriers, at velocities of such magnitude as to imply total destruction without advance warning." He stressed, however, that Americans should not despair. "Our country has faced agonizing tests more than once during its relatively short history. . . . It emerged each time from the crucible not without scars but with greater confidence and richer maturity." Those were stirring words from a German who, thirteen years earlier, was trying to design a rocket capable of hitting New York.

In his speech, von Braun provided four stark warnings about probable Soviet intentions. The rocket that carried Sputnik I and II aloft was, he maintained, "probably" one "originally designed to carry a thermonuclear warhead over intercontinental range." Second, that same rocket, with minor modifications, could put a payload of up to 100 pounds on the Moon. Or, thirdly, it could be used to put aloft spy satellites capable of keeping track of the "progress of all surface construction projects, ship movements, and air base operations anywhere in the world."[23] Finally, he warned, the Soviets already had in place a program for manned space travel. "A man in Sputnik means control of the globe," he told one newspaper.[24]

Von Braun appreciated that the best way to achieve his personal dreams of space travel was to manufacture a patriotic purpose for his rockets. In order to realize his selfish fantasies, he called for the peacetime equivalent of total war:

> The Soviet challenge is by no means restricted to military technology. It goes far beyond the realms of politics and armies. No longer is the task of coping with the Red menace the exclusive responsibility of generals and statesmen. The acid test involves every facet of our civilization, every part of our society: religion, economics, politics, science, technology, industry and education. Free men everywhere have been caught up in this grim competition. . . . What we are about to discover is whether a nation, who has rated its home run sluggers and its fullbacks above scientists and philosophers, can meet the total competition of aggressive communism, and still preserve its way of life.

The underlying message was that rockets were the only way to preserve the American dream. Von Braun called for an emotional commitment driven by what he termed "the will to supremacy." That will "is intangible, . . . it must come from the hearts and minds of our people, it cannot be legislated, budgeted or evoked by decree." Hitler might once have said the same thing. For those who might have been put off by that kind of message delivered in a German accent, von Braun ended with an appeal to a higher moral purpose that might be discovered in space. Echoing Hale and Tsiolkovsky, he promised:

> As we probe farther into the area beyond our sensible atmosphere, man will learn more about his environment; he will understand better the order and beauty of creation. He may then come to realize that war, as we know it, will avail him nothing but catastrophe. He may grasp the truth that there is something much bigger than his one little world.[25]

Eisenhower understood that von Braun was a trickster intent on leading Americans down the lunar path. In rebuttal, the president countered that massive expenditure on space was not necessary to safeguard national security. Unfortunately, in a contest determined by dramatic effect, the president never had a chance against von Braun, the spin-doctor for space. He was probably the only rocket man with a full-time publicist—and the only one who didn't need one. The public loved him. He spoke perfect English with just the right amount of foreign inflection to give his science credibility.* "He always seemed to be older than he was," remarked Paul Castenholz of Rocketdyne. "You looked up to him as having a lot of experience, a very strong vision. There was an aura about him that gave him credibility."[26] Like the devil, he had all the best lines. Around the time of Muttnik, a reporter asked: "What do you think American astronauts will find when they land on the moon?" "Russians," he replied.[27]

Eisenhower was still intent on preventing Americans from launching blindly into space. Rocket dogs, he insisted, were a distraction. The real issue was nuclear-tipped intercontinental ballistic missiles, something the United States was rather good at making. He did not want the

---

* It was widely assumed that brilliant scientists were supposed to sound foreign—a fact confirmed by Einstein and virtually every science fiction film.

resources needed to protect Americans from nuclear war to be diverted into pointless outer-space games. As he told a gathering of senior Republicans, he was "firmly of the opinion that the rule of reason had to be applied to these Space projects—that we couldn't pour unlimited funds into these costly projects where there was nothing of earthly value to the Nation's security. . . . [I] would rather have a good Redstone than be able to hit the Moon, for we [don't] have any enemies on the Moon!"[28]

Eisenhower's space woes were compounded by talk of a missile gap. The crisis was invented by the professional worriers at RAND and ruthlessly exploited by the Democrats. On November 12, 1957, in the wake of Sputnik, a National Intelligence Estimate forecast that the Soviets would have five hundred operational ICBMs by the end of 1962, while the United States would have only around sixty-five. Other estimates put the Soviet figure closer to one thousand. On the strength of this evidence, Eisenhower was lambasted for endangering the security of the United States, even though the American lead in bombers and total nuclear weapons was still huge. A delighted air force used the data to jack up its budget requests. In fact, the crisis was entirely bogus, as Herbert York, an adviser at the Pentagon and White House, soon discovered.

> When I . . . got to Washington . . . I had access essentially to everything about the Soviet Union that related to technology and military preparedness and so forth, I saw what other people didn't see. . . . I came to the view very soon that . . . the Soviet threat had been greatly exaggerated and it was still being greatly exaggerated. And it looked like it was going to continue to be greatly exaggerated."[29]

It subsequently transpired that monastery towers, grain silos, and even a Crimean War memorial had been counted as missile sites. Eisenhower knew all along that there was no crisis, but he could not reveal to anyone why he knew. Secret missions by U.S. spy planes had provided the president with very accurate estimates of Soviet strengths. But Eisenhower could not release this information because the missions were illegal and he had promised the Soviets that they were not happening. Not until well after the 1960 election was it revealed that the Soviets would have only 150 missiles by 1962. Later, that estimate was lowered to 50. Then, in the autumn of 1961, after a comprehensive analysis of satellite reconnaissance was completed, the USSR was discovered to have only four operational missiles, with another twenty under construction. By then, however, the damage was done, and a Democrat was in the White House.

On May 1, 1960, Francis Gary Powers, piloting a top secret U-2 spy plane, was shot down over the Soviet Union. Eisenhower at first denied engaging in missions of this type, but then was forced into an embarrassing admission after the Russians produced the pilot and the plane. What was not revealed until many years later was that the spy plane missions were shortly to be replaced by high-tech satellites. Just seven weeks after Powers was shot down, a GRAB (Gallactic Radiation and Background) satellite was launched. This was the first true spy satellite, a project that had been started within a year of Sputnik. Eisenhower could not boast of the achievement because he was still professing his desire to keep space free of military appliances. The launch does, however, give a significantly different twist to the president's refusal to share in the nation's panicked reaction to Sputnik. He was justifiably confident that progress was being made in the fields that mattered.

For most Americans, the "missile gap" and the space race were two sides of the same coin. The public was growing increasingly impatient with Eisenhower's leadership—or lack thereof. They demanded action, especially after the success of Explorer I was followed quickly by the embarrassing failure of a second Vanguard, and then by the crash of Explorer II. But no one had yet explained what the purpose of these rocket launches was supposed to be. The rocket mania was built upon fear— fear that if the Soviets controlled the high ground of space they would be able to control the Earth. But that fear was totally unfounded, as those in the know clearly understood. By this stage, however, there were too many vested interests in the space race, too many people who staked their careers on rocket dreams. No one was prepared to admit that the emperor had no clothes—that von Braun's great space venture had no identifiable purpose.

In place of goals, Americans were offered fantasies. Orbiting the Earth, they were told, would be the first step to exploring the cosmos. Someday man would travel to the Moon, the planets, even to distant galaxies. Space travel would lead inevitably to colonization. All this was held together by the flimsy mantra that man must explore. No less a figure than Major General John Medaris argued that "the overriding importance of space exploration takes on proper significance only if we

appreciate that it will make possible a new understanding of man's relationship with the infinity of Divine Creation. If we fail to read the
signs right, we may find progress pre-empted by the agents of the anti-
Christ with awful consequences."[30] It was, he argued, God's will that
Americans should lead the way in space.

On April 17, 1958, von Braun and Medaris told the House Committee on Science and Astronautics that the United States should develop the capability to move soldiers and materials quickly to "the scattered and fast-moving battlefields of the future."[31] Their proposal involved stuffing three or four soldiers into the nose cone of a rocket and,
boom, an instant army could arrive from the sky to put out a fire at
some remote trouble spot. The idea made no sense whatsoever, but,
coming from von Braun, it seemed to the rocket enthusiasts like gospel.
The platoon-in-a-can plan was ridiculed by the army, but that didn't
matter, since von Braun understood better than anyone that while rockets were wonderful, it was rocket men who stirred the juices of ordinary
people.

Von Braun also played up the idea that he was doing his bit to ensure the preservation of the human species. In a letter to Howard
Schomer, president of the Chicago Theological Seminary, he wrote:
"The material benefits to mankind which will accrue from expanded
physical frontiers will permit a greater number of *homo sapiens* to in-
habit the universe—will permit the survival even of this species when
our own solar system in some far distant eon is a collection of cold dead
rocks floating in the dark airless void—as it surely someday must be."
Von Braun argued that it was "profoundly important . . . that [we] travel
to other worlds, other galaxies; for it may be man's destiny to assure immortality not only of his race, but even of the life spark itself." The population explosion would, he predicted, eventually render life on Earth
intolerable. "If we remain earthbound with our fermenting population
already overflowing this Earth container, man can only survive through
ultimate totalitarian socialized government control of every aspect of
our birth, life and death."[32]

On March 17, 1958, on its third attempt, Vanguard finally made it to
outer space. Unfortunately, that success could never erase a reputation
for ridicule. That was a pity, since it was an impressive achievement—a
bona fide effort at space science. Vanguard was designed specifically for
research; it was not a surplus ICBM hastily rejigged for the purpose of
tossing something (anything) into the heavens. Its satellite was a so-

phisticated package of instruments capable of measuring the gravitational fields of the Earth and Sun, the solar wind, the pressure of sunlight, atmospheric conditions in outer space, and other important phenomena. All this was packed into a four-pound satellite that used the most advanced miniaturization technology and relied on power from newly developed solar cells. It was placed in an elliptical orbit with a maximum distance from Earth of 2,466 miles. Unlike Sputnik, which burned up in the atmosphere within a few weeks, Vanguard will remain in orbit for at least another thousand years. It was a massive achievement sadly doomed to share the ignominy of that first fiery failure and went largely unnoticed by the American public. Khrushchev called it a "grapefruit" and Americans were inclined to agree. Forget all that sophisticated science and engineering—the Russians had a dog in space!

The Vanguard effort would develop into a genuine, but unfortunately understated and underfunded program for space exploration. It was the precursor of the Viking, Mariner, Voyager, and Cassini space probes that would eventually travel to the outer planets, sending brilliant postcards home. Meanwhile, the army's program would morph into the manned space program, which would be showered with money in the 1960s in order to make sure that an American was the first to walk on the desolate landscape of the Moon.

In the brave new world of space travel, size clearly mattered, which is precisely why Vanguard was largely ignored. On May 15, 1958, the Soviets again underlined their supremacy in heavy lifting by launching Sputnik III, a behemoth weighing nearly 1,400 kilos. The new capsule, one Russian academician stressed, "could easily carry a man with a stock of food and supplementary equipment." Americans winced. "Even with no holds barred," von Braun told congressmen, "I think it will be well over five years before we can catch up with the Soviets' big rockets because they are not likely to sit idly by in the meantime." But, he told another group of legislators, even that time period was unrealistic since the Americans had yet to grasp the importance of space. "We are competing in spirit only," he sighed.[33]

What few people understood was just how limited the Soviet space effort was. They had a big rocket with enormous thrust that allowed them to place a variety of things into orbit. But the USSR was like the huge weight lifter who makes up in muscle what he lacks in brains. As long as the space race remained a lifting contest, the Russian lead would seem big. And as long as the American public was encouraged to be simplistic, they would perceive themselves inferior to their Cold War rival. But when missions grew more sophisticated and technology began to play a larger part, the Russian lead would be exposed as a sham. That lead was also exaggerated by Soviet secrecy. Only the successes were reported, the failures remaining secret for thirty years. In fact, they failed more often than they succeeded. On June 25, 1958, for instance, Korolev aimed his R-7 at the Moon, but the rocket failed miserably. No one heard of that debacle, but everyone heard when yet another Vanguard exploded on the very next day.

On April 2, 1958, before a joint session of Congress, Eisenhower called for the formation of a National Aeronautics and Space Administration (NASA). The new agency, which implied a huge investment in space research and development, was precisely what Eisenhower had wanted to avoid. But, in the hysteria that followed Sputnik, it was the least he could get away with offering. The new organization, built on the foundation of NACA, would have an overtly civilian identity. Twelve days after Eisenhower's speech, a bill establishing NASA was steered through the Senate by Lyndon Johnson and Senator Styles Bridges. The new agency was officially founded on October 1. In an overt attempt to underline the civilian nature of NASA and to assert his authority over von Braun, Eisenhower appointed Keith Glennan, president of the Case Institute of Technology, the first administrator of the new agency. The president had chosen a man in harmony with his own views about space. "Personally, I do not believe we can avoid competition in this field," Glennan wrote shortly after his appointment. "But I do believe we can and should establish the terms on which we are competing. We could thus place the 'space race' in proper perspective with all the other activities in the competition between the U.S. and USSR."[34]

The new administrator quickly found, however, that Congress was desperate for a race and had no scruples about throwing money into space. A typical attitude was that of James Fulton, ranking Republican on the House Space Committee: "How much money would you need to get us on a program that would make us even with Russia . . . and probably leap frog them?" he asked. "I want to be firstest with the mostest in space, and I just don't want to wait for years. How much money do we need to do it?"[35]

While NASA was taking shape, the National Security Council was formulating a strategy for space. Included in its report, released on August 18, 1958, was a timetable for space exploits. The United States, it suggested, should send probes and satellites to the Moon in the following year, with a view to an unmanned soft landing in 1960. Manned orbits of the Earth should occur in the same year. A manned circumlunar flight might occur sometime between 1962 and 1964, with a lunar landing in 1968. It was a highly ambitious schedule, but it nevertheless presumed that the Russians would reach every goal well before the Americans, and would land a man on the Moon in 1965. As for the point of it all, the report was somewhat vague. The United States should aim for "projects which, while having scientific or military value, are designed to achieve a favorable world-wide psychological impact."[36]

Glennan quickly commissioned Project Mercury, the first American manned spaceflight venture. A Space Task Group was formed at Langley Research Center in Virginia to oversee this venture, with Robert Gilruth in charge. He understood completely that his initial remit was "to put a man in space and bring him back in good shape—and do it before the Soviets."[37]

One of the first acts of the new space agency was to establish a public affairs office to promote NASA. In 1959, Glennan appointed Lieutenant Colonel John "Shorty" Powers to serve as the agency's liaison with press and public. For him it was an extension of what he had already been doing in the military, but in this case the product was much easier to sell.

The army was not pleased about being ordered to surrender the JPL and most of ABMA to the newly formed space agency. Without von Braun's rockets, the army seemed decidedly dull: a collection of infantrymen with rifles, some tanks, artillery, and some short-range missiles. Glennan later confessed that he "had not realized how much of a pet von Braun and his operation had become. He was [the army's] one avenue to fame in the space business."[38] But that was precisely what Eisenhower wanted to destroy. The army was keen on von Braun because he was an important weapon in the power game with the other two services. The president understood that the difficulties and delays so far encountered had resulted in part because the services were competing with one another to get into space. The only way to make space a national effort was to separate it from the military entirely.

Von Braun reacted to the formation of NASA like a boy whose ball had been taken from him. In his view, the only sensible course for the president was to turn the entire space program over to his army team, but of course Eisenhower had different ideas. On August 15, the day after the Senate confirmed the nomination of Glennan, von Braun fed a story to the Associated Press to the effect that he and his colleagues were inclined to go into private industry rather than have their work absorbed by NASA. Eisenhower simply ignored von Braun's petulance. He understood that the rocket man had a nose for money and therefore could not stay mad for long.

In private, Eisenhower confessed that he was "getting a little weary of von Braun and his publicity seeking."[39] This perhaps inspired the president's desire to score publicity points of his own. In July 1958, the White House let it be known that it wanted something really big to counter the impact made by Sputnik III. Project Score (Signal Communications by Orbiting Relay Equipment) involved broadcasting a recorded message from a satellite circling the Earth. Fearing failure, Eisenhower specifically forbade the usual prelaunch fanfare; the project would remain completely secret until it actually happened, and any leaks (it was understood) would immediately mean cancellation.

The original message was supposed to have been something suitably patriotic recorded by a Signal Corps officer. But when Eisenhower grew convinced that the project would succeed, he decided instead to substitute his voice broadcasting a Christmas message to the world.

> This is the President of the United States speaking. Through the marvels of scientific advance, my voice is coming to you from a satellite circling in outer space. My message is a simple one. Through this unique means I will convey to you and to all mankind America's wish for peace on earth and goodwill toward men everywhere.[40]

Everything worked perfectly. Virtually anywhere in the world a person with the right radio equipment could hear the president sending his good tidings for peace and goodwill. Taking a page from the Soviet's little book of propaganda, the Americans announced that they had placed in orbit a satellite weighing a hefty four tons. In fact, that was the combined weight of the 150-pound satellite and the entire upper stage of the rocket.

The press reaction was mixed. Some papers celebrated any American achievement in space, while others called it an elaborate and expensive publicity stunt. That is precisely what it was, but that same criticism could have been leveled at everything that had happened in space so far. The entire venture was little more than a high-tech circus. By joining the circus, Eisenhower demonstrated that he had lost the argument. He had, out of necessity, abandoned his calm pragmatism and had reluctantly entered a meaningless contest to score points.

The success of SCORE was completely overshadowed in the following month by Soviet exploits with their Lunik spacecraft. As the name implied, these aimed for the Moon. The first one, launched in January 1959, missed the Moon by over 3,000 miles. But then, Lunik II, launched on September 12, hit the bull's eye, crashing very near to the dead center. It was the first time a man-made object had ever reached another heavenly body. This suggested that the Russians, undeniably good at thrust, were also developing a skill for guidance. That fear was underlined three weeks later when, on the second anniversary of Sputnik, Lunik III looped around the Moon. It wasn't actually a lunar orbit, but rather a stretched elliptical orbit of the Earth which, by clever calculation, took it behind the Moon, where it took snapshots of the dark side. The photos had a predictable effect upon American self-confidence, demonstrating once again that Korolev had a keen sense of how to use his rockets to dent the American ego.

Eisenhower had desperately tried to avoid turning space into a race, but he had failed. At the end of 1959, even his own vice president began to make an overt appeal to the space enthusiasts. Nixon, like Johnson and Kennedy, had his eye on the White House, over which a full Moon shone. "Space," Nixon argued, "captures the imagination. It indicates power; the people do not downgrade the military potentiality of space. I would hope otherwise, but I do not think this is the case." In other words, the wishes of the people would be met, even if they were misguided. "If I thought Congress would support increased expenditures for medical programs, for foreign aid—dramatically larger—I would trade space for this, but they will not buy it."[41]

Nixon was probably right. The nation had convinced itself that it was falling dangerously behind the Russians. "How to Lose the Space Race!" *Newsweek* shrieked in October 1959. The author, Edwin Diamond, blamed the Eisenhower administration's limited vision and tight purse strings for the failure to respond to the Soviet challenge. "No amount of soft soap can gloss over the dismal fact: The U.S. is losing the race into space, and thus its predominance in the world."[42] Few people asked whether the race was even important. Norman Cousins, editor of the *Saturday Review* and a lonely voice of reason, remarked: "In the best of all possible worlds, we could be elated that man has unshackled himself from Earth gravity. Unfortunately we don't live in the best of all possible worlds."[43] Prestige, a thing which neither filled bellies, nor kept people warm, nor kept predators at bay, was suddenly the number one priority of an embattled nation. That was sad, but even more tragic was the fact that prestige was determined by criteria so base. Shortly after Sputnik, the American people took the road labeled shallow. Forget democracy, forget liberty, forget the American dream. The worth of a nation would henceforth be measured by its ability to put a dog in orbit.

Perhaps the saddest part of this era was the fact that those who argued that national prestige could be measured by one's position in the space race were actually right. Polls taken in Western Europe after Sputnik did show a marked decline in the esteem the United States inspired. In 1955, the percentage of Germans, or British, or French who thought the Soviets were militarily superior to the Americans was tiny. Five years later, those who thought the United States was the strongest hovered down around 20 percent. The implication to be drawn from these polls was that most Europeans felt that, over the long term, the Soviet Union would emerge as the dominant nation. Yet this was at a time when the American nuclear arsenal was eight times larger than that of her adversary and when Russian peasants fought over potatoes.[44]

The shallowness of the space race can be measured by three events that occurred in the last year of Eisenhower's presidency. On April 1, the United States launched Tiros 1, the world's first weather satellite. The photos sent back to Earth were stunning, and the satellite instantly transformed weather forecasting. (The United States, by the way, made the meteorological data available to anyone.) Twelve days later came Transit 1B, the world's first navigation satellite, which allowed ships to calculate their position with pinpoint accuracy. Both of these feats required highly precise launch and guidance techniques of which only the United States was capable.

In the following August, the Soviets launched Korabl Sputnik II, yet another big capsule which this time carried two dogs, Strelka and Belka, into orbit. On this occasion, the two pups, unlike poor Laika, were returned to Earth safely.

Guess which feat got the most attention.[*]

---

[*] In April 1960, the United States had ten satellites orbiting the Earth, while the USSR had two. A poll taken in Britain during that month asked which superpower had the most satellites. Only 7 percent of those polled got the answer right (*New York Times*, October 27, 1960).

7

Rocket Jocks

The 1958 World's Fair, held in Brussels, was seen by the Soviets as an opportunity to show off. Their pavilion, the largest of all, focused on the progress made since the Bolshevik Revolution. The centerpiece was a display showcasing full-size replicas of the first three Sputniks.

Slightly smaller was the American pavilion, which also celebrated progress, but of a different sort. Displays provided a window on the riches of ordinary American life—television, fashion, cosmetics, appliances, etc., etc., etc. Everywhere people were eating burgers, hot dogs, ice cream, and popcorn. Anyone worried about the complacency of American society and the recent Soviet coup in space (two sides of the same coin, according to the experts) would not have enjoyed the visit to Brussels.

But Americans are nothing if not resourceful. During the night, a team from the CIA, the Defense Intelligence Agency, and other espionage organizations assembled at a secret location, synchronized their watches, and descended upon the Soviet Pavilion. Being spies, they dressed in black, carried no identification, wore the latest decoder rings, and knew all the cool expressions like "terminate," "negative," "10-8," and "10-4." For three hours they pored over the Sputniks, taking hundreds of photos with miniature cameras. The technology that some Americans assumed had been stolen by the Soviets was stolen back.[*]

For nearly a year after the formation of NASA, Wernher von Braun was allowed to stew in his juices, while the new agency set out long-term plans for manned space flight. At first, he tolerated this arrangement, since he was busily working on a huge rocket called Saturn which satisfied his ambitions. But then the Defense Department decided that a rocket that size was not needed for ICBMs. Therefore, if Saturn was to

---

[*] The feat would be repeated in the early 1960s, when American spies diverted a Lunik moon explorer due to be exhibited at a trade fair in Mexico. The driver of the truck carrying the space probe was lured away from his precious cargo after being offered a night of hedonistic delight in a roadside motel—courtesy of Uncle Sam.

be built, its only logical home was NASA. Keith Glennan of NASA made it clear that he wanted the entire von Braun team, a move approved by Eisenhower on October 21, 1959. Von Braun continued to act like a spoiled child, but his resistance melted away once he realized that Glennan planned to give him the lion's share of NASA's booster systems and launch operations. John Medaris, who had always been a soldier first and foremost, could not, however, stomach seeing his command absorbed by a civilian agency. He therefore chose retirement, taking satisfaction in a career well spent. "No human being, without the guidance of the Lord, could have been right as much as I was," he later told *People* magazine.[1] Over 2,300 ABMA space specialists ignored Medaris and followed von Braun.

Just eight months after its foundation, NASA formulated what amounted to a mission statement. An internal committee chaired by Harry Goett set out a long-range plan in which manned space flight figured prominently. At its first meeting, members of the committee concurred that they "should not get bogged down with justifying the need for man in space in each of the steps but outrightly assume that he is needed inasmuch as the ultimate objective of space exploration is manned travel to and from other planets." The aerospace engineer Max Faget proposed that they should immediately set out a bold plan to reach the Moon, an idea supported by his colleague George Low, who felt that "this approach will be easier to sell."[2]

The Goett report deployed all the right buzzwords. The main goal was nobly expressed as the "expansion of human knowledge." NASA appreciated that it was important to preserve "the role of the United States as a leader in aeronautical and space science." The agency understood that its role was to explore "the problems involved in the utilization of aeronautical and space activities for peaceful and scientific purposes." That said, any discoveries of military relevance would be made "available to agencies directly concerned with national defense."

Everything, however, depended upon money. One proviso would echo through the next two decades: tight budgets had the effect of "arbitrarily limiting . . . activities to a narrow line and . . . greatly reducing the rate of approach to . . . long-term goals." With sufficient funds and commitment, "the manned exploration of the Moon and the nearby planets" would be "feasible," and "this exploration may thus be taken as a long-term goal of NASA activities." Translated from NASA-speak to plain language this meant: "We can do a lot if you give us the money, but cut our budget and there's no way we'll beat the Russians to the Moon."

There followed an interesting list of targets. Manned space flight would take place in 1961 or 1962. An unmanned vehicle would land on the Moon in 1963 or 1964. A manned circumlunar mission would take place between 1965 and 1967. Finally, an American would land on the Moon sometime "beyond 1970."[3] The Moon program would be called Project Apollo, a name first revealed in mid-1960. The name was dreamed up by Abe Silverstein, director of the Office of Space Flight Programs in Washington. "I thought the image of the god Apollo riding his chariot across the Sun gave the best representation of the grand scale of the proposed program," he explained.[4]

An ad in a comic book from the Mercury era touts the "Orbitprop," a "satellite on a string" consisting of two tiny astronauts "in authentic space travel outfits" enclosed in a clear plastic ball. The cost was $1.00, postage paid.[5] The toy was not a far cry from reality. Mercury, America's first experiment in manned space travel, was a disappointment to those raised on science fiction tales of flying saucers, or those who had followed closely the development of the X-series space planes. The capsule was essentially a nuclear warhead adapted for temporary human habitation. Mercury was 6 feet, 10 inches long by 6 feet, 2 inches in diameter at its widest point, and weighed 4,300 pounds. A container designed to hold a man, it was only slightly roomier than a coffin, an analogy no one liked to dwell upon. It was simply a tin can, something designed to be shot into space, to circle the Earth and then to come down again with a big splash. Mercury had no means of propulsion, aside from a few retro-rockets needed to get it out of orbit. The craft was cone shaped, tapering to a small round cylinder at the top. This gave the impression of aerodynamics, but that impression was misleading since drag and shape hardly mattered in the vacuum of space. The most important feature was not the tip, but the rump—the rounded bottom where the ablative heat shield and retro-rockets were located. The shield was designed to absorb and deflect the 3,000-degree heat of reentry. In other words, the craft would return to Earth bottom first.

It seemed simple, but building it was hugely complicated. In January 1959 the McDonnell Aircraft Corporation won the contract to produce twenty capsules for NASA. By late 1960, 13,000 people were involved in the construction. McDonnell ran three shifts, seven days a week. Around four thousand separate firms supplied materials, and McDonnell farmed out work to over two thousand separate subcontractors. A similar situation could be found where Atlas rockets were built at General Dynamics in San Diego and at Rocketdyne in Los Angeles. Eisenhower might not have liked the idea of a space race, but at the construction plants workers felt differently. At factories spread over twenty-five states, they welcomed the big contracts that would give them steady employment for the next few years. That's exactly the way NASA wanted it. Eventually, hundreds of thousands of workers scattered all over the United States would learn to love the space program for the simple reason that it put food on the table.

Building the capsule was one problem, choosing the men to put in it another. From the beginning, they were called astronauts, even though the word implies one who travels to the stars. The more appropriate term would have been cosmonauts, which confines them appropriately to the solar system, or cosmos. That term was preferred by Hugh Dryden, the deputy administrator of NASA, but he was overruled. "We regarded the American term . . . as rather boastful," the Russian Alexei Leonov recalled. "Privately, we used to joke that we were mere 'cosmonauts' not 'astronauts.'"[6]

NASA was looking for "a group of ordinary Supermen," but hadn't a clue where to look.[7] The qualities of an astronaut were not necessarily those of a pilot, even though that's the way it seems now. Back in the late 1950s, NASA was thinking only of reasonably intelligent, physically fit men who could be trusted to carry out orders. That made for a very long short list, which could include doctors, teachers, baseball players, and trapeze artists. NASA was about ready to open recruitment when Eisenhower intervened. He pointed out that, if the agency put up no rigid barriers to qualification, every fantasist who had ever read a comic book would apply. In addition, every member of Congress would put forward someone from his state, probably his own son. The president insisted, rather wisely, that an entirely arbitrary prerequisite be introduced, namely that the candidates had to have experience as test pilots. This meant that they would already be on the government payroll and, being military men, would be accustomed to following orders. Another advantage of test pilots was that they were undiluted patriots, steeped in Cold War dogma. They could be trusted not to question the rationale of a mission.

The test pilot requirement limited the field to some 580 names. Further stipulations regarding physical condition, length of service, flying time, and education reduced that number to about 110. They were called together and told that participation would be entirely voluntary and that no candidate need think that he would be jeopardizing his career if he refused. NASA expected that the vast majority would tell the agency where they could stuff their space capsule. But the lure of participating in this new venture outweighed the mundane nature of what the astronauts would actually be doing. Mercury was important not for what it was, but rather for where it might lead. This was supposed to be the dawning of a new age of space travel. Granted, it was not the kind of space travel the pioneers of the X-series planes might have envisaged, but it was revolutionary. Logic suggested that the little tin cans were the first step on a long voyage to the Moon and, from there, perhaps to Mars. The venture was also the closest thing to armed combat in a Cold War stubbornly short of military adventure. The elite group of pilots would go to war for their country on a new battlefield called outer space. How could they possibly refuse? Of the thirty-four men called to the Pentagon on February 2, 1959, and the additional thirty-five summoned a week later, only thirteen decided not to volunteer. NASA was so overwhelmed by the response that officials decided not to call the final group of forty-one men.

The reality of what they were doing and how NASA viewed them was driven home when they went through the selection procedure. The candidates' skills in piloting high-performance aircraft were unimportant. Instead, as if to underline that they were just a step above laboratory rats, they were given incredibly arduous and invasive physical examinations. NASA wanted individuals who would respond properly to the unique conditions of getting to and orbiting in outer space. That explains why there were a lot of psychologists on the training staff and why the candidates spent so much time in sadistic, stomach-churning contraptions spinning upside down while listening to a high-pitched wail, with a tube up their anus. The men chosen were those who demonstrated an ability to tolerate loneliness, sensory deprivation, motion sickness, claustrophobia, and doctors.

When asked about the examinations at a press conference later, John Glenn replied: "They went into every opening on the human body just as far as they could go. Which one do you think you would enjoy the least?" Every manner of blood, skin, stool, sperm, urine, and other medical sample was taken. "Nobody wanted to tell us what some of the stranger tests were for," Glenn recalled.[8] Psychologists subjected the astronauts to thirteen different "personality and motivation" tests and another twelve "intellectual functions and special aptitudes tests." In addition to the ubiquitous Rorschach Test, the list included Thematic Apperception, the Minnesota Multiphasic Personality Inventory, the Edwards Personal Preference Schedule, and Guilford-Zimmerman Spatial Visualization.[9] At times it seemed to the candidates that they were simply cheap specimens on whom any junior technician could try out his latest method of torture. On one occasion, they were taken to the city morgue to observe an autopsy of an old woman who had died of peritonitis. The smell was overpowering, causing a third of the group to vomit. "I tried to figure out just what the bleeding hell we were doing there," Michael Collins later wrote.* "Was this supposed to make us at home with death on Earth, in case we had to cope with death in space? Or were we really supposed to learn something here, from the awful obscene jumbled pile of poor old lady parts in front of us?"[10]

From the original five hundred, seven were eventually chosen—Gus Grissom, Gordon Cooper, John Glenn, Deke Slayton, Walter Schirra, Scott Carpenter, and Alan Shepard. They were introduced to the public on April 9, 1959, by Glennan:

> Ladies and gentleman, today we are introducing to you and to the world these seven men who have been selected to begin training for orbital space flight. These men, the nation's Project Mercury astronauts, are here after a long and perhaps unprecedented series of evaluations which told our medical consultants and scientists of their superb adaptability to their upcoming flight.[11]

No mention was made of their ability as pilots. What set them apart was that they were freaks of nature, super-fit men who could withstand conditions that would cause ordinary mortals to pass out, get dizzy, vomit, or burst into tears.

The men chosen were eerily similar, at least on the surface. All were firstborn sons, and all came from small towns. All had pretty wives. All had attended university, and five had earned degrees. Glenn had left college to join the Marines, an acceptable diversion, while Carpenter, the black sheep, had flunked out, a fact he somehow managed to keep

* Collins was in the third intake of astronauts, when this bizarre exercise took place.

from NASA's prying eyes. While they were often compared to Columbus, in truth the exploration part of their endeavor was not very important to them. They were rocket jocks—intelligent, able men, but not
particularly deep thinkers. What attracted them to the space program
was the possibility of going faster and farther than any man had gone
before. They were not particularly interested in what they might discover in space, what moved them was the journey itself. Of the seven,
only Carpenter showed an interest in astronomy and an occasional tendency to consider the deeper philosophical issues of man's purpose in
the cosmos. This caused doubt to arise; some NASA officials felt that his
"poetic" nature impeded his effectiveness.

One quality that was never talked about but was high on NASA's
list was marketability. The Seven were carefully chosen for public consumption. Their lives were thoroughly sanitized, with unpleasant aspects camouflaged. "The word at the time was that unless you were the
all-American boy, with a perfect family, you weren't going to fly,"
David Scott recalled.[12] They were endlessly coached on how to behave
in public and, indeed, during a flight. Part of their training included a
short spell at what the astronauts called "charm school." Collins recalled how NASA taught

> certain social skills . . . we came back to Washington for a couple of
> days and they told us that you're supposed to wear socks . . . that go
> on forever . . . [since] no one wants to see hairy legs. . . . And you had
> to hold your hands on your hips this way [palms down] and not that
> way [palms up] because people you don't want to talk about hold
> them the other way.

Even though they had the foul mouths typical of the test pilot profession,
they were expected to talk like altar boys when circling the Earth. Never
in the course of human history was the word "darn" uttered so often by
so many in the service of their country. The press cooperated fully in the
charade, ignoring the astronauts' fondness for strong drink and loose
women. "We were quite aware that the image that NASA was trying to
project was not quite honest," Walter Cronkite later admitted. "But at the
same time, there was a recognition that the nation needed new heroes."[13]

The importance of these heroic qualities was revealed at that very
first press conference, when these men, who came from a profession
noted for its solitude, had suddenly to formulate a public face. The man

who shined above the rest was Glenn, poster boy of the Mercury program. He dominated that first press conference by oozing down-home American qualities. Glenn already had a public face, as a result of a stint on *Name That Tune*, with the child star Eddie Hodges as his partner. He'd been invited on to the show in 1957 after completing the first coast-to-coast supersonic flight. He'd also, just for good measure, flown combat missions in World War II and Korea, in the process winning five DFCs. Especially when he had his picture perfect family in tow, he looked like a living re-creation of a Norman Rockwell painting. As the *Life* reporter John Dille later remarked: "John tries to behave as if every impressionable youngster in the country were watching him every moment of the day."[14] Johnson said that he was the kind of man "you want your boy to be."[15]

Glenn shone because he had a natural ability to sell himself. "Somehow, and without intending to, I found myself speaking for the group," Glenn later reflected.[16] None of the reporters asked questions hugely relevant to the task of being an astronaut, concentrating instead on family, religion, patriotism, and service. They wanted to know what the astronauts' children were up to, and what their wives thought about the idea of traveling in a rocket. Yet these men, aside perhaps from Glenn, had hardly given a thought to their families during the pursuit of their careers. Had family been all that important, they would not have been test pilots. Schirra, Shepard, Slayton, Carpenter, Cooper, and Grissom sat dumbfounded, desperately trying to come up with the right answer to these inane questions—the kind of answer NASA wanted. Glenn, meanwhile, coasted. "I don't think any of us could have really gone on with something like this if we didn't have pretty good backing at home," he replied. "My wife's attitude toward this has been the same as it has been all along through all my flying career. If it is what I want to do, she is behind it, and the kids are too, one hundred percent." Slayton, much more typical of the astronaut pool, replied with brutal honesty: "What I do is pretty much my business. . . . My wife goes along with it."[17]

"We didn't know what was going to happen," Glenn said of the early years of the space program. "We were making up the music as we went along." He, at least, had perfect harmony. The really annoying thing about Glenn, at least for his space colleagues, was that he was the genuine article, a man who wasn't actually faking it when he talked about God, religion, Mom, and the white picket fences of his Ohio home. He even revealed that he took religion "very seriously" and taught Sunday School. He had reporters eating out of his hands when he confessed: "I got on this project because it probably would be the nearest to Heaven I will ever get and I wanted to make the most of it." In contrast, when Schirra was asked about God and religion, he claimed that he had participated "actively" in "church activities," but then said: "I think I should like to dwell more on the faith in what we have called the machine age. We have faith in the space age. . . . All of us have faith in mechanical objects."[18] That was a very honest answer, even a profound one, but it wasn't the right answer—not the answer Americans wanted. They were obsessed with images of pioneers, conquering the West with a rifle and a Bible. That was why they loved John Glenn.

Toward the end of the press conference came a question no one wanted: "Could I ask for a show of hands of how many are confident that they will come back from outer space?" The astronauts had to put their hands up; otherwise they would have looked like idiots with a death wish. What the reporter was getting at, what he had in the back of his mind, and what was in the back of almost every American's mind was the way American rockets had a tendency to explode.

In raising their hands, the astronauts looked like an older version of what they were: model schoolboys who knew the answer to every question, who never annoyed the teacher, and who could be trusted with a hall pass. They had walked into a trap. America wanted to know if they were prepared to die in the great battle with the Soviets. Are you real heroes? Raise your hands. They did so, felt like idiots doing so, but woke the next morning adored by all of America. All they had done so far was endure tubes up their asses, getting whizzed around in the medical version of a fairground ride, and, finally a press conference. But in a country desperate for heroes, that was enough. John Glenn raised both hands.

The press conference was NASA's first serious attempt to present its image to the public. It succeeded beyond its wildest expectations. Space had been presented not as a technical problem to be tackled by scientists and engineers, but rather as a human problem to be conquered by heroes. Even though in the grand scheme of things the astronauts would actually be unimportant (initially, at least), they were introduced to the public as the most crucial elements in the space equation. James Reston of the *New York Times* could hardly contain himself: "What made them so exciting was not that they said anything new but that they said all the old things with such fierce convictions. . . . They spoke of 'duty' and 'faith' and 'country' like Walt Whitman's pioneers. . . . This is a pretty cynical town, but nobody went away from these young men scoffing at their courage and idealism."[19]

Glenn, who more than anyone embodied the human side of the space program, eventually came to regret the excessive attention paid to personality. After his mission in 1962, he complained that the public should be more interested in the scientific aspects of his trip than with his wife's hairstyle. John Kennedy, by then president, did not agree. "My own feeling is that [both] are equally important, in the sense that we are proud of this trip because of its scientific achievement and we are also proud of it because of the men and women that are involved. Our boosters may not be as large as some others, but our men and women are."[20]

Space, the new frontier, would be conquered by young men embodying old values. The qualities that America would take into space would be the same as those that had driven the pioneers westward. Speaking to a script, Carpenter described space as a "fabulous frontier," while Shepard said that the desire to be a pioneer was what motivated his decision to join the program. Grissom likewise explained that he was attracted to the "spirit of pioneering and adventure" and said that, had he lived in the previous century, he would probably have been an explorer of the American West.[21] The press swallowed the pioneer image like a fresh slice of Johnnycake. There were constant references to Columbus, Magellan, Captain Robert Scott, and Charles Lindbergh. The infinite nature of space provided limitless room for overblown hyperbole. According to Reston, the astronauts' exploits would make "Columbus and Vasco de Gama look like shut-ins . . . and their exploration may open up more in the heavens than the old sailors did in the sea."[22]

The unknown would be vanquished by the familiar. Americans delighted in the way the romantic conception of the warrior hero had been resurrected from the ashes and dust of the post-atomic landscape. The atom bomb, it once seemed, had vaporized the romance of war; it was the ultimate assertion of technology over man. But, suddenly, a new version of the medieval knight had emerged. As *Newsweek* remarked, "The drama of the human spirit—solitary, vulnerable, curious—facing the unknown elements of the universe is as old as mankind."[23] In the space race, the two Cold War adversaries would meet in vicarious conflict; they would challenge each other in an artificial contest—totally trivial but hugely momentous.

It was all one great fantasy, one collective willingness to indulge a romantic illusion. Americans didn't really want to confront the stark realities of what they were attempting. They didn't want to think of the cost. They didn't want to think about whether the quest had point, purpose, or value. Not long after that initial press conference, some reporters ran into Chuck Yeager, the brave uncle of the Magnificent Seven. He was asked whether he regretted not being given the chance to be an astronaut. He bluntly replied that he had no regrets since he'd been able to fly some of the most sophisticated aircraft ever developed. In contrast, he remarked, the Mercury Seven would not really be flying. Yeager might as well have said that Lindbergh couldn't pilot a bus. The reporters stood aghast, barely able to take in what was being said. Sensing their confusion, Yeager politely pointed out that, if the seven were indeed pilots, why was a chimp going to take the first flight? He had presented the space version of the Emperor's New Clothes, voicing what everyone knew, but no one wanted to contemplate. The wire services decided not to print the interview.

The men at Edwards Air Force Base who piloted the rocket planes felt that they had been forgotten, put in a file marked "redundant programs." They saw themselves as the true pioneers; their planes were the forerunners of real spacecraft, not just tin cans thrown into orbit and brought down by a computer when a techno-geek on the ground so decided. They couldn't understand why everyone considered Mercury astronauts heroes, even though they had never actually flown one of those capsules. But none of that mattered. Mercury had been so adroitly sold to the American public that only a few insiders cared about what was going on at Edwards. An inaccurate distinction had been made between planes and rockets, and NASA had successfully convinced the public that it had a monopoly on the latter. The guys at Edwards were victims of Cold War politics, men left behind in the shallow race for results. The line going around Edwards went "Shepard, Grissom and Glenn, the link between monkey and man."[24] In truth, however, Yeager and his pals weren't laughing.

The astronauts were, as Yeager liked to say, simply Spam in a can. They were an exclusive group doing a decidedly mundane job. Propulsion, guidance, altitude, speed, and reentry would all be controlled by technicians sitting behind consoles on the ground. The astronauts would have some control over yaw, pitch, and roll by way of hydrogen peroxide thrusters, but this wasn't real flying, at least not to someone who'd flown a rocket plane across the United States at twice the speed of sound. They'd have only slightly more control over their craft than the average airline passenger who can adjust his seat back and tray table and pull down the window blind. "Mercury was basically an unmanned system," Slayton admitted. "It was designed to operate unmanned."[25]

The Mercury engineers, either out of ignorance or spite, did their best to demean the pilots. The first capsule design didn't include a window, on the grounds that, in purely utilitarian terms, the astronaut had no need to know where he was going since he had no control over his direction. Nor, upon landing, could they slide back the canopy and jump out while looking incredibly manly, as jet pilots loved to do. The capsule was so cramped that the astronaut had to be helped in by a team of handlers, who then put the door in place and bolted it shut. In the early days, some engineers even boasted that Mercury was so high tech that the astronaut was completely redundant, someone along for the ride—about as important as the hood ornament on a Cadillac. Serious consideration was given to drugging the astronauts for the entire journey, not because the effects would be particularly painful or nauseating (which they might), but rather to ensure that these intrusive overachievers wouldn't be tempted to press buttons or turn knobs just to see what might happen. Good sense eventually prevailed, however, and the realization dawned that an astronaut asleep would not present the right kind of image to the public. People might ask why he had to be there at all. Besides, the chatter of an astronaut would provide wonderful background noise to the otherwise silent journey in space. And the one thing astronauts could do that computers could not was smile.

The astronauts rebelled at their redundant status. They wanted changes which, at the very least, provided the suggestion of control. At first, the training consisted of them learning how to ride in a capsule, not how to fly it. But they argued that no system was foolproof, and that they should therefore be the backup navigation system in case something went wrong. They also insisted on a window, since everyone knows that you can't drive a vehicle without seeing where you're going. Then they demanded a hatch that could be opened by the astronaut himself, in other words something to show that NASA had a bit more faith in them than it had in the chimps. Finally, they asked NASA to quit using the word "capsule," since no one ever pilots a capsule. They wanted it to be called a spacecraft. NASA officials tried hard to make the adjustment, but "capsule" never completely disappeared.

Slayton, the most combative of the original seven, did not appreciate Yeager's thoughts on space flight, even if they had been kept out of the papers. In October 1959, he addressed the Society of Experimental Test Pilots conference at the Beverly Hilton. In the room were a number of men who were thinking of applying for the astronaut program. In a pointed reference to Yeager, he announced that he was going to talk about the requirements "for the pilot, or astronaut, in Project Mercury and follow-up programs." Heavy emphasis was placed on the word "pilot." "Objections to the *pilot*," Deke Slayton remarked, "range from the engineer, who semi-seriously notes that all problems of Mercury would be tremendously simplified if we didn't have to worry about the bloody astronauts, to the military man who wonders whether a college-trained chimpanzee or the village idiot might not do as well in space as an experienced test *pilot*." The military man's opinion, he maintained, was hardly worthy of serious consideration. As for the engineer, "The answer . . . is obvious and simple. If you eliminate the astronaut, you concede that man has no place in space." The speech played cleverly to the crowd, but, perhaps unintentionally, Slayton gave credence to the engineer's position. He went on to discuss what the "pilot" of the Mercury capsule would do, a list of tasks embarrassingly modest, and admitted that "if everything works perfectly the pilot's task will be quite simple."[26] He provided no proof that the pilot was actually essential to the mission and could only justify his presence by repeating the vague gospel that man belonged in space. But for the vast majority of the audience, that was enough. They agreed with Slayton that man had a place in space, even if his presence made the task more complicated, more expensive, and more limited in potential.

The day after the big press conference, Scott Carpenter's wife, Rene, woke up to find reporters parked in her front yard. She invited them in and then took her children to school. When she returned, she found that they had rearranged her furniture in order to convert her living room into a studio. "It's as if I'd been acting on a dark stage all my life," she later remarked, "and suddenly someone turns on the spotlight."[27] Walt Bonney, a NASA public affairs officer and the man who had engineered that first press conference, understood immediately that the Mercury astronauts were instant celebrities and that this would have enormous impact on their lives. He realized that they would need a handler to steer them through the celebrity maze. He got in touch with Leo De-

Orsey, a tax lawyer extraordinaire who represented a fair number of Hollywood celebrities. DeOrsey agreed to take on their portfolio pro bono. He understood that they could not be constantly tripping over journalists on their voyage to the Moon. So he proposed selling exclusive rights to a single highest bidder.

The auction was opened, but in fact there was ever only one logical winner, namely *Life* magazine. The style of that magazine, and the dominance of photographs within, fitted perfectly with the wholesome image DeOrsey wanted to project. "I believe *Life* was far more effective than NASA in bringing national attention to the men themselves," Dora Hamblin, one of the journalists assigned to cover the space program, remarked. "The magazine's circulation was about 8 million at the time, and the *Life* stories were far more detailed, more comprehensible, than most of the NASA releases. *Life* was very good . . . at clarification of complex subjects. Our pictures were far superior to anything NASA did— so much so that NASA customarily picked up *Life* photos for service to other news media."[28]

The magazine bid $500,000 for a three-year contract giving them exclusive rights to the astronauts' stories (and those of their wives), to be divided equally between the seven. That worked out at $23,809.52 per astronaut per year, about triple what they were making in basic pay from the military. The first feature came out on September 14 and was headlined "Ready to Make History." A week later, the wives of the seven were on the cover under the headline "Seven Brave Women Behind the Astronauts."[29] From 1959 to 1963, over seventy articles were published in the magazine about the astronauts or their families, and on twelve occasions they or their wives appeared on the cover.[30] *Life*, rather appropriately, also provided life insurance for the astronauts, since they could not get coverage from any private company. Thanks to the magazine, the astronauts had become both rich and famous even though none of them had yet gone up in a rocket.

"*Life* treated the men and their families with kid gloves," Hamblin later confessed.

> So did most of the rest of the press. These guys were heroes, most of them were very smooth, canny operators with all of the press. They felt that they had to live up to a public image of good clean all-American guys, and NASA knocked itself out to preserve that image. I had to clear all of the first-person stories through NASA public relations in

Washington—after I had previously cleared them with the men themselves. Thus there was a tendency to keep everything . . . "peaches and
cream."[31]

In the Soviet Union, Alexei Leonov enjoyed reading *Life* with all its detail on the astronauts' lives. "I knew all their names and how many children they had. There were endless photographs of them with their families relaxing in the comfort of their spacious condominiums and smart
houses." Long after the fact, he discovered how much *Life* had paid the
astronauts, and found this intriguing. "Our salaries at that time were
around 50 or 60 rubles a month, and that was before the deduction of
Communist Party dues. There was no official exchange rate . . . but on
the black market a dollar fetched about 60 kopecks; so our salaries
would have been the equivalent of about $100 a month."[32]

The *Life* contract was not unanimously well received. A short time
after Kennedy took office in 1961, his press secretary, Pierre Salinger, announced that, in line with the "ethical standards" policy of the new administration, public servants would not be able to receive additional
compensation in the form of writing articles or making public appear-
ances. This meant that the *Life* contract would not be renewed. A few
weeks later, however, during a weekend spent together water skiing,
Kennedy asked Glenn what he thought about the commonly voiced
criticism that a soldier in combat, say a Marine at Iwo Jima, faced the
same risk of death as an astronaut, yet did not expect recompense from
Time, Inc. Glenn replied that the risks were indeed the same, but that
the soldier at Iwo Jima did not have to live in a fish bowl, with his every
action watched by the public, with cameras pointed and questions
thrust at him, his wife, his children, his milkman, and his dog. Kennedy,
who knew what it felt like to be famous, apparently agreed. Within
weeks, it was quietly announced that the "ethical standards" policy did
not apply to NASA.[33]

The *New York Times* nevertheless queried the right of astronauts "to
reap enormous private profits" from what was supposed to be a great
patriotic effort. *Newsweek* likewise questioned "how much a hero can
expect to gain financially and still remain a hero." It ridiculed how *Life*
had turned the space program into a "Barnumesque extravaganza"
with the astronauts and their families resembling "cardboard characters of soap operas."[34] Granted, a lot of this could be placed in a file labeled "sour grapes." "I think you have to honestly say that it hasn't hurt

the agency," Tom Paine, who took over as NASA administrator in 1968, remarked of the contract. "But I must also tell you that I would never have signed such a contract, on the principle that I don't think government employees, who are paid by the public, should have done this."[35]

The *Life* money went a long way in Cocoa Beach, the town closest to the Cape Canaveral base. Cocoa defined shabby: nothing ever seemed to be new; paint was always cracked and shutters hung from one hinge. That's what the salt air and the oppressive sunshine did. Since back then few people came to Cocoa for a holiday, no one saw the need to make it look pretty. Everyone was there for a purpose, and that purpose, nine times out of ten, had something to do with space. Cocoa Beach grew as the space adventure grew, all the while adding more of those cheap houses which looked old the day after they were finished. The town, surely one of the ugliest in America, could easily be mistaken for one gigantic strip mall.

The astronauts, however, didn't find Cocoa particularly unpleasant. Nor, for that matter, did they mind Houston, where Mission Control was situated, another city that thrived on rocket fuel. For them, both places were terra cognita. Anyone who'd spent their career in the military was familiar with ugly towns and cheap housing. The only thing different about Cocoa was perhaps the weather and, for most of the year, that was a plus. It sure beat Edwards Air Force Base, which was like homesteading in an oven.

In a display on the Mercury years at the Astronaut Hall of Fame in Cape Canaveral, one finds a short article probably inspired by the PR gurus at NASA. It describes the typical astronaut. He has a pretty wife, three kids, a dog, drives a Corvette, is 5 feet, 10 inches tall, weighs about 165 pounds, and has a crew cut. But all that was meant for public consumption. No mention was made of the fact that some astronauts cheated on their wives, often in serial fashion, and sometimes with more than one woman at a time. They liked to drive their Corvettes very fast, and to see if the cars could do things not recommended in the manual. In truth, they didn't all have Corvettes. Schirra had a Maserati (which he bought from Brigitte Bardot) and Carpenter had a Shelby Cobra. Glenn, rather strangely for a man who wore his patriotism like an overcoat, drove a German-made 583 cc Prinz. The gas mileage was great. His other car was a station wagon.

In the springtime, millions of little black bugs would hatch in the swamps surrounding the space center in Houston. Their distinguishing feature was that they mate in midair. Grissom once told a reporter that he deeply envied those bugs. "They do the two things I like best in life," he confessed, "flying and fucking—and they do them at the same time."[36] According to an oft-told tale, Grissom once left a party to take his wife to the airport and then returned with another lovely babe he'd found there. It was also widely rumored that he had fathered a child out of wedlock, after docking with a secretary at McDonnell.

"It was so wide open in those days," Paul Haney, a NASA public affairs specialist, confessed. "I can recall seeing lines—three or four ladies in a line—outside the Oceanside rooms of the good old boys."[37] Several space groupies boasted about their carnal knowledge of six astronauts, but no woman ever collected a full set—Glenn remained forever faithful. One of the women who managed an almost-full house was a Florida reporter named Wicky—her astronaut lovers were inevitably dubbed the Wicky Mouse Club. After working her way through the first group, she started on the new recruits, eventually bagging a few who would walk on the Moon. One astronaut later confessed: "If they only knew that it wasn't all that difficult, then they probably wouldn't have bragged that much."[38]

Glenn worried that his rutting colleagues would undermine the clean image essential to selling space. On a number of occasions, he lectured his fellow astronauts on the need to show prudence. They told him, as politely as possible, to mind his own business. After all, this was the era in which the president of the United States was a serial womanizer who cheated on his mistresses. Yet the press obligingly presented an image of a harmonious First Family. The same whitewash was applied to the astronauts. "I knew . . . about some very shaky marriages, some womanizing, some drinking and never reported it," Hamblin confessed. "The guys wouldn't have let me, and neither would NASA. It was common knowledge that some marriages hung together only because the men were afraid NASA would disapprove of divorce and take them off flights."[39] Everyone knew the truth, but no one told it.

If the astronauts were stereotyped heroes, their wives were stereotyped American wives, or at least that's how they were presented. Their makeup was always perfect; they wore the latest fashions; they were model mothers; they were beautiful without being sexy; they knew how to make perfect chocolate chip cookies. All of them had the ability to turn on a 200-watt American smile, which suggested that they were perfectly happy with the lives their husbands (and NASA) had chosen for them. They also knew how to look reverentially at their husbands, the new gods of space. Underneath that veneer of beauty and confidence, however, anxiety lurked and suffering ran deep. They worried about their spouses, not just about the danger of what they were doing, but also about the ever-present temptations they found irresistible. Some turned to alcohol. Some demanded divorce. But, in most cases, the couples stayed together so that the boys could get their flights.

The Mercury astronauts were presented to the world as clean, wholesome heroes—the Eagle Scout who grows up but remains uncorrupted. It was, of course, all bunk. What is interesting about the creation of heroes is how willingly the press participated in the charade and how trusting the public was—as if everyone had decided to be gullible. But that gullibility fitted an age, just as lifting the lid on the astronauts suited the age of cynicism that followed. In the 1960s, naiveté was not just a state of mind, it was a policy. America, a nation at war, craved comic book heroes.

The popular media loved the Magnificent Seven, but some specialists who knew about rockets and space weren't so sure. They questioned the hype that surrounded Mercury and, more importantly, expressed doubts about the scientific merit of the program. To them, the whole thing seemed like a stunt designed solely to get an American into space before a Russian. "After all the noise and excitement of the great space exploit have died away," Lee DuBridge of Caltech remarked in 1962, "we must still face up to the question of what it's all about. Aside from national pride, what is the space program really for?"[40] In 1958, the then NACA director Hugh Dryden told the House Space Committee that the idea of putting a man in a capsule and shooting him into space atop a Redstone rocket "has about the same technical value as the circus stunt of shooting a young lady from a cannon."[41] In so doing, he ruled himself out of consideration for the top job at NASA. On September 9, 1960, Keith Glennan assured reporters that Mercury was part of a long-term program of space exploration and entirely independent of what the Russians had done, or might do, in space. The fact that he had to make this statement cast doubt on its veracity.

On July 29, 1960, the Magnificent Seven went to Canaveral as guests of honor at the first test of a Mercury-Atlas vehicle. According to the plan, the first astronaut would go aloft in a suborbital flight aboard a Redstone rocket, with the more powerful Atlas reserved for the first orbital flight. Hundreds of VIPs attended, and journalists swarmed like wasps. NASA should have known better. Memories of Vanguard had apparently faded. The countdown was suitably dramatic and the blast-off absolutely superb. The rocket went straight up and then gradually arced over on a seemingly horizontal trajectory, as rockets do. And then it blew up. It didn't just fizzle, or slow down or break into pieces. It blew up. Flaming debris shot in all directions. For about a minute, it rained rocket offshore. Everyone who witnessed the spectacle was suddenly reminded of that first press conference, when the astronauts were asked to raise their hands if they thought they would survive the space adventure.

NASA invited everyone back for another party later in the year, with the showpiece this time being the entire Mercury/Redstone package—everything but the astronaut. Congressmen and senators were given the red carpet treatment and promised the best fireworks display they'd ever seen. Then came the countdown. The announcer said "zero" and then "ignition," but nothing happened. No "lift-off." No noise. There was a little shudder, a bit of smoke, and, then, rather like a popgun, the cap of the rocket shot off. It went 4,000 feet into the air and then a parachute opened and this little thing drifted innocently back to Earth amid a crowd of silent observers all with their mouths wide open. As a demonstration of Mercury's emergency escape procedure, it was absolutely brilliant, but that's not what it was supposed to be. The Redstone just stood there like the little rocket that couldn't.

The people at NASA, from the minor technicians right through to the Magnificent Seven, were all deeply worried. The manned space program had so far created an enormous amount of publicity and expectation, but nothing of substance. Some of them were worried that a new president, elected in November 1960, might simply decide that Mercury was tainted with Eisenhower's reputation for failure, and scrap the whole program. George Low, who at that time was working on preliminary studies for a manned lunar landing program, later confessed that he was not very certain where the man-in-space idea was headed. "We did not know whether the country would support a major effort beyond Mercury in manned space flight."[42]

In a memo to Abe Silverstein of the Office of Space Flight Programs dated October 17, 1960, Low advocated the bold approach. "It has become increasingly apparent that a preliminary program for manned lunar landings should be formulated. This is necessary in order to provide a proper justification for Apollo, and to place Apollo schedules and technical plans on a firmer foundation. As he later explained,

> We were . . . quite concerned, of course, that in the subsequent year's budget, which was being prepared at that time, there were insufficient funds for any major lunar program. And we felt it would be most important to have something in the files, to be prepared to move out with a bigger program should there be a sudden change of heart within the Government, within the administration, as to what should happen.[43]

A change of heart by Eisenhower was hardly likely. A change of government was.

Senator John Kennedy, his cross-hairs trained on the White House, made good use of the Redstone and Atlas failures, touting them as symbols of Eisenhower's ineffectual leadership. A friendly journalist described the Massachusetts senator as "a clean cut, smiling American boy, trustworthy, loyal, brave, clean and reverent, boldly facing up to the challenges of the Atomic age."[44] To many, he seemed the perfect candidate to lead the American team in the space race. "I look up and see the Soviet flag on the Moon," he taunted the Republican candidate Richard Nixon during their debate. "Polls on our prestige and influence around the world have shown such a sharp drop that up till now the State Department has been unwilling to release them."[45] Reiterating the complacency theme, he taunted Nixon: "You yourself said to Khrushchev, 'You may be ahead of us in rocket thrust, but we're ahead of you in color television.'" Personally, he retorted, "I will take my television in black and white. I want to be ahead of them in rocket thrust."[46] The space issue was perfectly suited to a candidate who presented himself as a man of action. It provided a direct link between Washington and Hollywood. Kennedy was Buck Rogers in a business suit.

The truth was radically different. A month after the fiasco at the Cape, an event occurred that might have vindicated Eisenhower's space policy, if only the news could have been released. On August 24, the Discoverer 14 spy satellite sent its precious cargo of photos back to Earth. These were recovered in midair by an air force C119 Flying Boxcar—the first successful recovery after numerous failed attempts. When the photos were processed, they revealed 1.5 million square miles of Soviet territory in astonishingly accurate detail. Among the discoveries were sixty-four airfields, twenty-six surface-to-air missile sites, and a major rocket launch facility. One little satellite, in less than a week, had collected more information on the enemy than all the U-2 flights had collected in four years of highly risky operations. More importantly, the feat demonstrated that the United States was way ahead of the Soviet Union in the space techniques that mattered. That is what it should have demonstrated, had Eisenhower been allowed to boast. Unfortunately, Discoverer had to remain a secret.

"The people of the world respect achievement," Kennedy proclaimed in Portland, Oregon, during the presidential campaign.

> For most of the twentieth century they admired American science and American education, which was second to none. But they are not at all certain about which way the future lies. The first vehicle in outer space was called Sputnik, not Vanguard. The first country to place its national emblem on the moon was the Soviet Union, not the United States. The first canine passengers in space who safely returned were named Strelka and Belka, not Rover or Fido, or even Checkers.*[47]

What Kennedy failed to mention (because it was not in his interests to do so) was that the United States was already ahead in the space race. In the three years since Sputnik, America had successfully launched twenty-six satellites and two space probes, while the figures for the USSR were six and two.

Elections are seldom won on facts. If not for Sputnik, it is difficult to see how Kennedy, or for that matter the Democrats, could have won in 1960. Kennedy was a lot like NASA. On the surface, both were handsome, articulate, bold, and brave. Underneath, both were manipulative, mendacious, scheming, and untrustworthy. Nevertheless, in terms of the implications for space, Kennedy's election was probably not that significant, since both he and Nixon (and Johnson, for that matter) were prepared to pander to the populist desire to race in space. But, in contrast to the era of Eisenhower, the advent of Kennedy was momentous, perhaps a watershed. The age of reason was over. The age of artifice had begun.

* Checkers was a cocker spaniel that saved Richard Nixon's political career in 1952. Nixon, the vice presidential candidate, was accused of setting up a secret slush fund. On television, he said the only illegal gift he'd ever received from political cronies was Checkers. Since his children loved the dog, he wasn't going to give it back—even if it was a crime. America bought the story.

8

Before This Decade Is Out

After all the attention given to space in his campaign, Kennedy hardly mentioned the matter in his inaugural address. He concentrated instead on foreign affairs, rousing the American people with a call to arms, which, through brilliant writing, managed to sound like a noble crusade. The only mention of space came halfway through the speech when Kennedy expressed his regret at the way the Cold War had divided the world. He offered the Soviets some vague areas in which cooperation and trust might be fostered. "Let both sides seek to invoke the wonders of science instead of its terrors," he urged. "Together let us explore the stars, conquer the deserts, eradicate disease, tap the ocean depths, and encourage the arts and commerce."[1]

For the nervous types at NASA, and many were nervous in January 1961, that statement fell far short of what they had expected from the new president. He had, after all, relentlessly hammered Eisenhower over the space issue, therefore it was reasonable to expect that he would set out bold new plans. What exactly did "explore the stars" mean? That could be done with a telescope from Sears. And what did he mean by "together"? Was the president seriously thinking that the United States might cooperate with the Soviets? What was the point in that?

When NASA submitted its 1962 financial year proposals to the Bureau of Budget in May 1960, Eisenhower discovered, supposedly for the first time, that the agency had long-term plans for a lunar landing. Somewhat worried by this revelation, he asked his science adviser George Kistiakowsky to critique NASA's manned spaceflight proposals. Kistiakowsky, who worried that Mercury would be "the most expensive funeral a man has ever had,"[2] formed a six-man committee, headed by Brown University chemistry professor Donald Hornig. The report, completed on December 16, 1960, made sobering reading. Hornig and his colleagues showed no symptoms of rocket fever.

We have been plunged into a race for the conquest of outer space. As a reason for this undertaking some look to the new and exciting scientific discoveries which are certain to be made. Others feel the challenge to transport man beyond frontiers he scarcely dared dream about until now. But at present the most impelling reason for our effort has been the international political situation which demands that we demonstrate our technological capabilities if we are to maintain our position of leadership. For all of these reasons we have embarked on a complex and costly adventure.[3]

Mercury was described as a "somewhat marginal effort, limited by the thrust of the Atlas booster." Like Kistiakowsky, the committee was not impressed. "The fact that the thrust . . . is barely sufficient for the purpose means that it is difficult to achieve a high probability of a successful flight while also providing adequate safety for the Astronaut." There followed a warning about the dangers of embarking on ambitious programs for the wrong reasons. "A difficult decision will soon be necessary as to when or whether a manned flight should be launched. The chief justification for pushing Project Mercury on the present time scale lies in the political desire either to be the first nation to send a man into orbit, or at least to be a close second."

The report went on to discuss Apollo. Improvements to the Saturn booster would allow "successively more elliptical orbits which carry it further and further from the Earth, culminating about 1970 in a manned flight around the Moon and back to the Earth." As regards landing on the Moon, the committee concluded that Apollo did not show much promise, but it did perhaps offer the opportunity of developing skills necessary for such a landing. In the meantime, however, "it should be possible to obtain far more detailed information about the Moon by unmanned spacecraft and lunar landing craft than the crew of the circumlunar flight could gain."

A lunar landing would be "the first really big achievement of the man-in-space program." But was there any point in it all? The answer Eisenhower really wanted came later in the report, when the committee addressed the issue of whether demonstrable advantages could be gained from putting men into space.

Certainly among the major reasons for attending the manned exploration of space are emotional compulsions and national aspirations.

These are not subjects which can be discussed on technical grounds. However, it can be asked whether the presence of a man adds to the variety or quality of the observations which can be made from unmanned vehicles, in short whether there is a scientific justification to include man in space vehicles.

By that criterion alone, the program fell short. Putting a man in space complicated and limited the venture. Two crucial points followed: "The degree of reliability that can be accepted in the entire mechanism is very much less for unmanned than for manned vehicles. As the systems become more complex this may make a decisive difference in what one dares to undertake." In addition, "From a purely scientific point of view it should be noted that unmanned flights to a given objective can be undertaken much earlier. Hence repeated observations, changes of objectives and the learning by experience are more feasible." These points should have been obvious to anyone thinking rationally about space, but rocket fever had overtaken rational thought.

The conclusions were devastating: "It seems, therefore . . . that man-in-space cannot be justified on purely scientific grounds. . . . On the other hand, it may be argued that much of the motivation and drive for the scientific exploration of space is derived from the dream of man's getting into space himself."[4] In other words, ego was more important than science. The cost of an ego trip around the Moon, the committee thought, would be at least $8 billion. When Eisenhower and his close advisers discussed the report, an aide recalled that the mood was "almost sheer bewilderment—or certainly amusement—that anybody would consider such an undertaking. Somebody said, 'This won't satisfy everybody. When they finish this they'll want to go to the planets.' There was a lot of laughter at that thought."[5] The report was sufficient to convince Eisenhower to veto Apollo. He explained that he was not prepared to "hock his jewels" in order to send a man to the Moon.*

Among those who agreed with Eisenhower was Glennan, the NASA administrator. "If we fail to place a man on the Moon before twenty years from now," he confided to the president in October, "there is nothing lost." The president thought Glennan was "screwing up his courage to state publicly that this should not be done."[6] Eisenhower's veto of Apollo

---

* He was alluding, rather inappropriately, to the myth that Isabella had sold her jewels in order to pay for the voyage of Columbus. In fact, she had done no such thing.

was the last act of a president desperately trying to keep a technocratic monster at bay. His was a quixotic struggle, but no less noble for its futility. In his farewell address, he warned, famously, that, "in the councils of government, we must guard against the acquisition of unwarranted influence, whether sought or unsought, by the military-industrial complex. The potential for the disastrous rise of misplaced power exists and will persist." That warning was, to say the least, prophetic. But allied to that danger was the power wielded by men like von Braun and Teller:

> Akin to, and largely responsible for the sweeping changes in our industrial-military posture, has been the technological revolution during recent decades. In this revolution, research has become central, it also becomes more formalized, complex, and costly. A steadily increasing share is conducted for, by, or at the direction of, the Federal government. . . .
>
> Yet, in holding scientific research and discovery in respect, as we should, we must also be alert to the equal and opposite danger that public policy could itself become the captive of a scientific-technological elite.[7]

Speaking to a reporter the next day, Eisenhower bemoaned the "insidious penetration of our own minds that the only thing this country is engaged in is weaponry and missiles." The words were all the more powerful coming from a man who had depended so heavily on armaments throughout his professional career. "The reason we have them is to protect the great values in which we believe," he reminded his listener.[8]

A new president commissioned a new committee. Kennedy's transition team on space was headed by Professor Jerome Weisner of MIT. His committee drank at the same well of skepticism as had Hornig. Its report, submitted on January 10, 1961, began by drawing attention to how the enthusiasm for space travel had moved attention away from the most important rocket issue, namely that of maintaining America's nuclear defenses. As regards the matter of manned space flight, the Weisner committee acknowledged that there was a certain inevitability to the venture, more because of esoteric notions than practical advantages:

> Space exploration and exploits have captured the imagination of the peoples of the world. During the next few years the prestige of the

United States will in part be determined by the leadership we demonstrate in space activities. It is within this context that we must consider
man in space. Given time, a desire, considerable innovation, and sufficient effort and money, man can eventually explore our solar system.
Given his enormous curiosity about the universe in which he lives and
his compelling urge to go where no one has ever been before, this will
be done.

The committee nevertheless perceived something seriously wrong with
America's space priorities. The great danger facing the administration
was that it might get pulled into an enormously costly venture whose
only justification was its popularity. A warning followed: "Space activities are so unbelievably expensive and people working in this field are
so imaginative that the space program could easily grow to cost many
more billions of dollars per year."

The area singled out for praise was that of scientific investigation.
"In spite of the limitations in our capability of lifting heavy payloads,
we now hold a position of leadership in space science." The committee
maintained that the United States should build on this lead, perhaps by
developing techniques for the "discovery of extra-terrestrial forms of
life" which would be "one of the greatest human achievements of all
times." In this field, however, the committee had in mind experiments
that could be conducted by "instruments," not man. That was where
the U.S. lead was most profound, but maintaining that lead would not
be easy. "The USSR has a number of competent scientists. It will be easier for them to catch up with us in instrument development than for our
engineers to catch up with the Russians in the technique of propulsion.
Thus we must push forward in space science as effectively and as forcefully as we can." That task would prove all the more difficult if too
much attention was given to the macho "space race."

The committee felt that the emphasis upon superficialities was primarily a problem of leadership at NASA, which was dominated by engineers rather than scientists. A manifestation of this problem could be
found in the man-in-space program, which the agency presented as an
end in itself, rather than as a means to achieve great scientific break-
throughs. The committee accepted that "manned exploration of space
will certainly come to pass and . . . that the United States must play a
vigorous role in this venture." But, like the Hornig group, the Weisner
committee felt that manned space travel was a potentially dangerous

and expensive distraction. "A crash program aimed at placing a man into an orbit at the earliest possible time cannot be justified solely on scientific or technical grounds. Indeed, it may hinder the development of our scientific and technical program . . . by diverting manpower, vehicles and funds." Furthermore, the enormous enthusiasm generated by the Mercury program had

> strengthened the popular belief that man in space is the most important aim of our non-military space effort. The manner in which this program has been publicized in our press has further crystallized such belief. It exaggerates the value of that aspect of space activity where we are less likely to achieve success, and discounts those aspects in which we have already achieved great success and will probably reap further successes in the future.

There followed a stern warning: "A failure in our first attempt to place a man into orbit, resulting in the death of an astronaut, would create a situation of serious national embarrassment. An even more serious situation would result if we fail to safely recover a man from orbit." In the starkest terms, Kennedy was warned that Mercury was taking on a momentum of its own and might prove impossible to stop.

> By allowing the present MERCURY program to continue unchanged for more than a very few months, the new Administration would effectively endorse this program and take the blame for its possible failures. . . . If our present man-in-space program appears unsound, we must be prepared to modify it drastically or even to cancel it. . . .
>
> Whatever we decide to actually do about the man-in-space program, we should stop advertising MERCURY as our major objective in space activities. Indeed, we should make an effort to diminish the significance of this program to its proper proportion before the public, both at home and abroad. We should find effective means to make people appreciate the cultural, public service and military importance of space activities other than space travel.[9]

In other words, the space program (rather like Vietnam) was a quagmire threatening to engulf the new administration. The committee did not mention that the mess was partially of Kennedy's making, since he had encouraged the American people to draw dangerous conclusions about the significance of Sputnik. The most glaring problem with the space program was the absence of a coherent and rational purpose. The American people had been led to believe that putting a man in space was important, yet to pursue that goal might actually result in the neglect of other, far more important concerns, with a consequent weakening of national security. Kennedy had been given some highly pertinent advice. But whether he could act on this advice was open to doubt. Like Dr. Frankenstein, he had created a monster, and he had to live with it.

Kennedy's first task was to appoint a new NASA administrator, a difficult job since he hadn't yet formulated his space policy. Weisner was pushing him in one direction—that of good sense—while the public, cheered on by von Braun, was pulling him in another. He asked James Webb, a man who had no scientific background but knew government as well as anyone, to take the job. Webb was reluctant, but refusing a president wasn't easy. He took the job and named Hugh Dryden his deputy and Robert Seamans associate administrator. Both provided the technical expertise that Webb lacked.

Webb was a shrewd political operator whose experience in Washington stretched back to New Deal days. A lawyer by training, he'd worked for the Kerr-McGee Oil Company and McDonnell Aircraft, in the process learning about big business and big government, while collecting powerful friends. He'd been budget director under Truman, so he knew precisely how the system of allocating federal funds worked. "I'd call him a politician," Brian Duff, director of NASA's Public Affairs Division, reflected, "except it sounds like I'm diminishing him, and I'm not diminishing him. He was a politician in the best sense." Duff recalled a revealing Webb habit:

> Whenever he talked to the White House, he'd face the White House. He'd stand at his desk when he was on the phone with the White House and he would face the White House. When he talked to the Hill, he turned around and faced the Hill. I don't think he knew he was doing it, but he would physically turn and look in the direction.[10]

The one-time gamekeeper used his knowledge well in his new job as poacher. One of the things he understood fully was that the space program had to be deeply imbedded into the American way of life. This was not just a case of getting people excited about the adventure. It was much more fundamental than that: the agency would have to become a combination of benevolent uncle and Mafia boss, overseeing the interests of the people, and rendering those people indebted to NASA. Contracts would be distributed across the fifty states, with no area neglected. The agency would be an employer of hundreds of thousands of people. Others would see their children go to university on NASA-funded programs. For others still, the agency would be seen as the inspiration for new inventions that improved the quality of life. As *Time* observed, NASA under Webb "sprouted like Jack's beanstalk, sucking up men and money at a prodigious rate, sending its tendrils into every state."[11]

Everything depended on Congress remaining friendly. That meant that money had to flow to the right areas, and that news had to be carefully managed. It also meant that members of Congress had to be supplied with the right sort of information, so that they could make favorable judgments when NASA appropriations came up for discussion. Agency officials received extensive coaching from the public affairs office in order to prepare them for testimony before Congress. Politicians were supplied with extensive "justification books" that provided detailed breakdowns of how the money would be spent and what would result from it. By 1963, these books were already running to four volumes.[12]

Within NASA, there was deep uncertainty about what the new administration would mean, despite all of Kennedy's rocket talk during the campaign. For the first few months of his administration, Kennedy displayed no great enthusiasm for space, in line with Weisner's caution. Reporters began to doubt his commitment. On February 8, 1961, he told the press: "We are very concerned that we do not put a man in space in order to gain some additional prestige and have a man take disproportionate risk, so we are going to be extremely careful in our work." He tried to downplay the importance of the race: "Even if we should come in second in putting a man in space, I will still be satisfied if when we finally do put a man in space his chances of survival are as high as I think that they must be."[13]

This must have stuck in Webb's gullet. An empire builder and a showman, he did not agree to take on NASA in order to see it relegated to a minor civil agency overseeing studiously cautious short steps into space. On March 21, 1961, he told Kennedy that the space program offered "no better means to reinforce our old alliances and build new ones." Furthermore, an ambitious role in space would bring about new areas of international cooperation in meteorology and communications. . . . The extent to which we are leaders in space science and technology will in large measure determine the extent to which we, as a nation, pioneering on a new frontier, will be in a position to develop the emerging world forces and make it the basis for new concepts and applications in education, communications, and transportation, looking toward more viable political, social, and economic systems for nations willing to work with us in the years ahead.[14]

The stirring message was designed specifically for a president who prided himself on his vision and who spoke enthusiastically of new frontiers. Webb's sales pitch was the left jab, and Lloyd Berkner's subsequent report the right hook. At the end of March 1961, his National Academy of Sciences (NAS) Space Science Board came up with a series of upbeat reasons why the United States had to be ambitious. Berkner, a close friend of Webb's, argued that "scientific exploration of the Moon and planets should be clearly stated as the ultimate objective of the U.S. space program for the foreseeable future."

In direct contrast to the Hornig and Weisner recommendations, the Berkner group "strongly emphasized that planning for scientific exploration of the Moon and planets must at once be developed on the premise that man will be included." The group seemed in little doubt that valuable scientific research was possible only if man were present:

> From a scientific standpoint, there seems little room for dissent that man's participation in the exploration of the Moon and planets will be essential . . . Man can contribute critical elements of scientific judgment and discrimination in conducting the scientific exploration of these bodies which can never be fully supplied by his instruments, however complex and sophisticated they may become.

Then came the stirring stuff, carefully written to harmonize with a president who had spoken of new frontiers in his inaugural address:

> The members of the Board as individuals regard man's exploration of the Moon and planets as potentially the greatest inspirational venture of this century and one in which the entire world can share; inherent here are great and fundamental philosophical and spiritual values

which find a response in man's questing spirit and his intellectual self-realization.[15]

The NAS report was supposed to be scientific, yet here were scientists pushing the spiritual advantages of a manned program. In fact, the report was a stitch-up engineered by Berkner to serve his friends in NASA and in the White House. Committee meetings exhibited a great deal more disagreement than Berkner's summary ever indicated. Some members later regretted giving him excessive freedom to determine what went into the report. He had delivered little more than pro-NASA propaganda, designed to suit Webb's purposes, and had even assured Webb, a month before the report was released, that recommendations would be favorable.[16]

The real importance of this report lay not in its message, for ideas of this sort had been expressed many times. The importance lay in its timing. The Kennedy administration had been buffeted by political storms in Berlin, Cuba, and Indochina during the first two months of Kennedy's presidency and was therefore rather short of good news. Events had yet to harmonize with the crusading zeal expressed at his inaugural. Berkner had presented Kennedy a diversion from other problems and a way to rouse the people. The reception given to the Magnificent Seven had been enormous. Their popularity showed no sign of waning, thanks in part to the clever PR people at NASA and the heart-warming space stories appearing regularly in *Life.* One of the greatest attractions of the space issue was that it was nonpartisan; it allowed Kennedy to appeal to parts of the country he could not ordinarily reach. Given the prevailing mood in the country, the temptation was great to ignore Weisner and hitch his wagon to the Mercury capsule.

NASA was meanwhile forging ahead with Mercury. According to the plan, the first passengers were pigs. They were strapped into the capsules and dropped from great heights in order to test Mercury's resistance to high-impact landings. Though the pigs suffered some minor internal injuries, the resilience of the capsule impressed the engineers—proof that pigs could fly.

Next came the chimps. They were trained at a specially constructed facility at Holloman Air Force Base, part of the White Sands missile range in New Mexico. Chimps, as everyone knows, are smarter than dogs. These ones were clever enough to learn that positive reinforcement is not all that great if the task in question is rather unpleasant. Their trainers therefore quickly figured out that the only way to get the reluctant primates to do what was wanted was through "operant conditioning," a.k.a. negative reinforcement. If they failed to cooperate with a test, they were given an electric shock through the soles of their feet. They were presented to the American public as hairy, smiling space travelers, rather like a Buck Rogers version of Cheetah. In actual fact, those smiles were simian versions of a snarl. Every chance they got, they took the opportunity to wreak vengeance on their tormentors.

Toward the end of 1960, the chimps were moved to Cape Canaveral. In a similar manner to their human colleagues, they had undergone a selection process in order to find the best of the bunch. The original group of forty was whittled down, first to eighteen, then to six. Four females and two males took the trip out to the Cape. Ahead of them awaited the distinction of being the first American (not to mention primate) to be taken to space in a rocket. Up to this point, NASA had resisted giving the chimps names—this project was, after all, about science, not sentimentality. But the publicists decided that Ape No. 61, a male chosen to be first in space, needed a moniker so that NASA could show off its human side. On the day the flight was announced to the press, he was christened Ham, an acronym for Holloman Aerospace Medical Center. *Life* featured the "astrochimp" on its front cover. The human astronauts didn't much enjoy sharing the limelight with an ape, especially one who implied that what they were doing wasn't exactly rocket science. In his darker moments, Shepard hoped that the first Mercury test would result in a "chimp barbecue."[17]

Launch day, January 31, must have seemed to Ham like any other day. He was woken up, fed, probed, prodded, and then stuffed into his cubicle, which was then depressurized. Ahead, it seemed, lay another day of getting bounced around—low level torture punctuated by the occasional electric shock. But then the cubicle was taken up an elevator and placed inside a bigger container—the Mercury capsule. And then, as would become commonplace in the space program, everyone waited while a number of preflight glitches were corrected. The only one who wasn't anxious, it appears, was Ham himself. He was quite happy to wait, since for the first time in his ape memory no one was zapping his toes.

That was the calm, and then came the storm. When the rocket finally took off, it climbed at a steeper angle than planned, which massively increased the g-forces bearing down on Ham. Within seconds, he felt like he was lying under a two-ton pile of coconuts. But the chimp performed like a champ. Even though his heart went into overdrive, he remained reasonably calm. Before long, the g-forces quit and he experienced something entirely new—weightlessness. Though that was strange, it was infinitely preferable to being squashed. Ham sat back and did as he was taught. When a white light came on, he had fifteen seconds to push a lever at his right hand. If he delayed a second too long, he got shocked. That was easy. But then came the blue light, which lit up every two minutes. When it came on, he had only five seconds to pull the lever at his left hand or else he got zinged. Positive reinforcement was not completely absent. NASA had rigged up a machine which dispensed banana pellets when lights and levers were deployed in correct sequence.

Everything went brilliantly. Ham seemed oblivious to the momentous nature of what he was doing. Then the retro-rockets fired, the g-forces increased again, and the capsule hurtled back toward Earth. The rockets were unfortunately jettisoned earlier than intended, which meant that reentry was 1400 miles per hour faster than it was supposed to be, increasing the force to 15gs. All these problems meant that the capsule landed 130 miles off target, and the navy took two hours to find it. Again, Ham took it like a man, even though a man had never done what he'd done. But he wasn't very happy. When the navy divers finally opened the hatch, all he wanted to do was bite someone anywhere. But perhaps the worst part of the whole journey came when he was paraded before the press and the flashbulbs sent him into a rage. That was not a lesson NASA expected, but it was a lesson nonetheless.

According to the plan, the next flight, scheduled for March 24, was supposed to take Alan Shepard on a short, suborbital flight. Ham's mission seemed to indicate that the venture was safe and NASA was ready. Robert Gilruth, head of the Mercury program, certainly thought so, and so did Shepard. But Wernher von Braun didn't like the way the Redstone had performed. Ham had been given a much wilder ride than had been intended. Von Braun was also undoubtedly aware of the worries expressed by Weisner. If Shepard died, that would be the end of the space program. Von Braun got his way. The next test went up without man or chimp. Shepard was furious at being denied the opportunity to be the first man in space. "We had 'em," he wrote many years later when the bitterness had still not subsided. "We had 'em by the short hairs and we gave it away."[18]

American caution meant that the first man in space was named Yuri Gagarin. Korolev (the Russian chief engineer), aware that the Americans were gaining on him, decided it was time for another space spectacular. On April 12, 1961, Gagarin was stuffed into a Vostok capsule and shot into orbit. Shortly after 4:00 A.M., the news reached Washington. A reporter, thinking it sufficiently serious, phoned "Shorty" Powers, the NASA press liaison, explained what had happened, and asked for a comment. A bewildered Powers muttered, "We're all asleep down here." Collecting his thoughts, he then managed to mutter congratulations to the Russians.[19] But the damage had been done. The headline writers had an easy day.

"Let the capitalist countries catch up to our country," Gagarin said shortly after his momentous flight.[20] Some Americans tried to be civil; others simply wallowed in self-pity. "The fact of the Soviet space feat must be faced for what it is," said the *Washington Post,* "and it is a psychological victory of the first magnitude for the Soviet Union."[21] The *New York Times* pointed out that JFK "had attempted to present himself as . . . a young, active, and vigorous leader of a strong and advancing nation." Unfortunately, however, "none of Kennedy's achievements . . . have had the effectiveness or the spectacular quality of Soviet efforts."[22]

Some commentators took the opportunity to point out that, whatever the Soviet Union might achieve in space, it was still a heathen empire. Gagarin unwittingly played to that tune. When asked whether his travel through the heavens had altered his feelings about a supreme being, he replied: "I do not believe in God, I do not believe in talismans, superstitions and such things." This was, Americans would point out, proof that "all . . . notions of a religious relaxation in the Soviet Union must undergo a shocking revision."[23] The Russians thereafter delighted in fueling a theological fire. "I don't believe in God," Gherman Titov, the second Russian in space, said; "I believe in man—his strength, his possibilities and his reason. . . . Sometimes the people say God is living [in space]. But I haven't found anyone there."[24] If they were simply trying to rile the Americans, they succeeded brilliantly. "Much as we admire the Russian achievement in space travel," Reverend Dr. Ralph Sockman of New York remarked in a sermon, "we almost have to smile at the childishness of Soviet scientists who assert that the failure to find angels and spirits in heavenly space tends to strengthen disbelief in God."[25]

Gagarin was Kennedy's Sputnik. Prestige was again pushed to the fore. Across Europe polls revealed that the number of those who thought the Soviets were technologically superior and militarily stronger than the United States was steadily increasing. Meanwhile, newspapers in Africa, Asia, and the Middle East showered the accomplishment in hyperbole. According to which paper one read, the achievement was more important than the invention of the printing press, the discovery of the New World, or the advent of the wheel.[26]

For the American people, Kennedy's reaction was eerily reminiscent of Eisenhower. "I do not regard the first man in space as a sign of the weakening of the . . . free world," he argued. "But I do regard the total mobilization of men and things for the service of the communist bloc over the last years as a source of great danger to us. And I would say we're going to have to live with that danger and hazard for much of the rest of this century." When journalists asked whether he was tired of being in second place, he replied: "However tired anybody may be, and no one is more tired than I am, it is a fact that . . . the news will be worse before it is better, and it will be some time before we catch up. We are, I hope, going to go in other areas where we can be first, and which will bring perhaps more long-range benefits to mankind."[27]

Those remarks were, of course, a dish specially cooked for journalists. In private, Kennedy did not try to hide his dismay. "Is there any place we can catch them?" he asked a small group of experts that included Webb, Weisner, and Dryden, who had gathered at the White House. Weisner counseled caution, as would be expected. "Now is not the time to make mistakes," he said. The budget director, David Bell, was worried about the cost. But Kennedy was determined to push forward. "When we know more, I can decide if it's worth it or not. If somebody can just tell me how to catch up. Let's find somebody, anybody, I don't care if it's the janitor over there, if he knows how. There's nothing more important."[28]

The United States was not behind at all, at least not in any way that mattered to her security. The fact that Gagarin had orbited the Earth did not make the Soviet threat any greater. But Premier Nikita Khrushchev liked to suggest that it did, and he was a master of bluff. He encouraged ordinary Americans to believe that Gagarin's orbit implied that the Soviet Union had a huge arsenal of ICBMs ready to rain down on American cities. They had nothing of the sort. Due to American advantages in satellite technology, the White House knew how things stood, but it could not reveal what it knew. As a result, the American people concluded that they were desperately behind the Soviets. In fact, in the areas that really counted, they were five years ahead.

Kennedy was doing everything he could to resist falling into the black hole of space. But then, on April 17, 1961, just five days after Gagarin's fantastic flight, the United States sent a puppet army of disgruntled Cuban exiles into a place called Bahia de los Cochinos. The world would know it as Bay of Pigs. According to the CIA scenario, the "superbly trained" soldiers would cut through the commies and spark a popular revolution that would topple Fidel Castro. Things didn't quite work out that way. The operation was crushed, leaving Kennedy deeply embarrassed. It seemed that the United States could do nothing right. This suggestion was reinforced when, a few days later, on April 25, another test of the Atlas rocket ended disastrously. The rocket went astray and had to be blown up by remote control after just 40 seconds. A second test, on the 28th, lasted only 33 seconds before explosion again painted the sky.

Gagarin rekindled all the feelings of inferiority that had surfaced after Sputnik I. Within Congress, the mood was tense impatience. "I am darned tired of being second," Overton Brooks, the Democrat from Louisiana, complained. "My objective is to beat the Russians."[29] Gagarin's flight took place just as the House Committee on Science and Astronautics was reviewing NASA's FY1962 budget. James Fulton, the Pennsylvania Republican, demanded that NASA start working "around the clock." He told Webb: "Tell us how much money you need and we on this committee will authorize all you need."[30] Victor Anfuso, a Democrat from New York, shouted: "I want to see our country mobilized to a wartime basis because we are at war. I want to see our schedules cut in half. I want to see what NASA says it is going to do in 10 years done in 5."[31] In the atmosphere of impatience and alarm there was little opportunity to examine critically the justification for man in space. Webb walked away from his hearings delighted at the experience of finding a committee willing to give him more money than he wanted. As Weisner quipped, "Just wave space at Congress and they give you a billion dollars."[32]

Congress was ready to run, but Kennedy still refused to take a lead. His lack of action, particularly after the Gagarin flight, provoked serious questioning of his leadership. At a press conference on April 21, a reporter pointedly remarked: "You don't seem to be pushing the Space Program nearly as energetically now as you suggested during the campaign." Kennedy replied with lots of statistics to suggest that great progress was being made, and then went on to remind journalists that a space program would be enormously expensive. The influence of Weisner was unmistakable. "We have to consider," he added, "whether there is any program now, regardless of its cost, which offers us hopes of being pioneers in a project. It is possible to spend billions of dollars in these projects in Space to the detriment of other programs and still not be successful."

Kennedy was no longer a candidate running for president who could make any irresponsible promise he liked. He was now president, and had to pay attention to budgetary constraints and the return an investment might yield. "Now I don't want to start spending the kind of money that I am talking about without making a determination based on careful scientific judgments as to whether a real success can be achieved or whether because we are so far behind now in this particular race we are going to be second in this decade." In other words, being behind in space wasn't quite the disaster he once suggested. He added: "I don't think we ought to rush into it . . . until we really know where we are going to end up." As for the possibility of beating the Russians to the Moon, Kennedy answered: "If we can get to the Moon before the Russians, we should. . . . [But] we first have to make a judgment, based on the best information we can get, whether we can be ahead of the Russians to the Moon."[33]

Weisner was still counseling caution. According to Gilruth, he was holding up the first Mercury flight because doctors had been telling him that a human being might not be able to stand weightlessness. Weisner wanted a long series of chimp flights before Mercury could be man-rated. Since NASA had been doing weightlessness training for some time, one suspects that the medical objection was just Weisner's way of trying to kill a program he thought dubious. Then came Gagarin's flight, which cleared the medical obstacle and seemed to demand an American response. There was no room left for caution.

While Kennedy was urging journalists to be patient, the ground was being prepared for a major space initiative. On April 20, he asked Johnson to conduct a study of capabilities, in order to find a contest

America might win. "Do we have a chance of beating the Soviets by putting a laboratory in space, or by a trip around the Moon, or by a rocket to land on the Moon, or by a rocket to go to the Moon and back with a man?" he asked in desperation. "Is there any other space program which promises dramatic results in which we could win?"[34]

Glen Wilson, who spent nineteen years as a professional staffer on the Senate Space Committee, feels that the phrasing of Kennedy's request was hugely significant: "The very first question y'know: what can we do. To. Beat. The. Russians!? He didn't say: What can we do to advance the scientific effort here; what can we find out if we send scientists to the Moon? He didn't say any of that stuff. He said: what can we do to beat the Russians?"[35]

Johnson consulted a loaded deck of "experts." Scientists were conspicuous by their absence. The vice president did not hear the honest opinion of the Space Sciences Board, because Berkner muzzled those members who felt a lunar mission would be an expensive drain on "real" science. Von Braun, in contrast, was all for going to the Moon, and suggested that all other priorities be put "on the back burner" in order to achieve that goal. One of LBJ's business cronies, Donald Cook, voiced the old argument about prestige, but never quite so overtly. He advised that future action had to be "based on the fundamental premise that achievements in space are equated by other nations in the world with technical proficiency and industrial strength . . . and will be of fundamental importance as to which group, the East or the West, they will cast their lot."[36]

Enthusiastic support came from a somewhat unexpected quarter. Robert McNamara, the defense secretary, was intent on applying sophisticated management techniques and cost-benefit analysis to the defense sector. The aim was to reduce budgets and to squeeze a greater return out of every dollar. Such a goal would not appear to harmonize with an open-ended commitment to NASA, which burned money quicker than rocket fuel. But NASA was a *civilian* agency, a crucial distinction. By supporting NASA, McNamara ensured that vast sums were funneled into the aerospace industry without affecting the defense budget. The industry giants, including Lockheed, General Dynamics and Douglas, were in deep financial trouble, crying out for government help. McNamara liked the idea that they would be rescued by a Moon mission and the costs of rescuing them would never show up on his ledgers. According to Weisner, the opportunity to rescue the aerospace industry "took away all argument against the space program" as far as Kennedy was concerned.[37]

Johnson's answer came back in the form of a three-page memo dated April 28. First came the prestige argument: "Other nations, regardless of their appreciation of our idealistic values, will tend to align themselves with the country which they believe will be the world leader—the winner in the long run. Dramatic accomplishments in space are being increasingly identified as a major indicator of world leadership." Time was running out, Johnson warned. "If we do not make the strong effort now, the time will soon be reached when the margin of control over space and over men's minds through space accomplishments will have swung so far on the Russian side that we will not be able to catch up, let alone assume leadership." On the question of beating the Russians, Johnson argued that the only chance for the United States lay in a race to the Moon:

> Neither the U.S. nor the USSR has such capability at this time, so far as we know. The Russians have had more experience with large boosters and with flights of dogs and man. Hence they might be conceded a time advantage in circumnavigation of the moon and also in a manned trip to the moon. However, with a strong effort, the United States could conceivably be first in those two accomplishments by 1966 or 1967.[38]

There it was, stated simply. If the United States wanted to regain world prestige lost to the Russians because of their achievements in space, it should embark now on a trip to the moon. Here was a race that Americans might just be able to win.

A space program was also attractive to Kennedy and Johnson for reasons entirely distinct from space or the Cold War. Both men were interested in social regeneration, and both were keen on New Deal–style work programs that would encourage economic growth and benefit the less well-off in society. But Kennedy had won the election by a slender margin and was dependent upon fiscally conservative southern Democrats who did not want to spend huge sums on people's problems. Space was an opportunity to hide a massive domestic spending program within a Cold War Trojan Horse. Conservative Democrats who might otherwise have been inclined to reject such a program would support it if they felt it essential to the struggle against communism.

Two botched Atlas launches should not have affected Shepard's flight, since he would be riding a Redstone. Nevertheless, it was difficult to be optimistic when it seemed that almost everything the Americans sent up from Cape Canaveral did a short manic dance in the sky before blowing to smithereens. It is fair to say, therefore, that a sizable proportion of the hundreds of thousands of onlookers who crowded the roads, fields, and beaches in the vicinity of the launch site on May 5, 1961, expected to witness a disaster.

Shepard, who had practiced the full routine of the short flight in around 120 separate simulations, was confident that every element in the procedure was dependable, except, that is, for the most important one, the performance of the rocket itself. The one thing he did not practice, however, was how to deal with delay—specifically, what to do if he was overcome by a call of nature while waiting for the guys on the ground to push the button. Those guys were determined not to let the flight go ahead until every variable had been checked, then rechecked and then checked again. Meanwhile, Shepard, stuffed into his tiny tin can, sensed the smallest little signal from his bladder. The delay stretched on, and the signal grew more persistent. Eventually his bladder seemed bigger than the capsule itself. He thought he might explode. Leaving aside the obvious embarrassment of the first American astronaut wetting himself prior to his launch, he was worried that free-flowing urine would short circuit systems and abort the flight. Or it might confuse the cooling in his pressure suit, suddenly transforming it into a refrigerator. Or worse: he might cause a spark that would turn the pure oxygen atmosphere into an inferno. Eventually, all those considerations were rendered immaterial; he simply had to pee. He informed the Capcom of his intention. The message came back: permission denied.

A full bladder waits for no man. Eventually Shepard realized that there were limits to how heroic he could be. When the countdown stopped for what seemed the hundredth time, he demanded permission to empty. This was reluctantly granted. Unfortunately, since he was tilted backwards in the capsule, with his bottom slightly higher than his shoulders, the tug of gravity meant that the urine traveled inexorably toward his neck, shorting out sensors along the way. But that was nothing compared to the pain he had previously felt.

He was wet, but he was ready. The guys on the ground, however, weren't happy about the pressure in the rocket, which seemed too high. Shepard, listening in on their deliberations, grew increasingly impatient. He knew that, if they decided the problem was serious, the flight would be delayed for at least a couple of days. Having wet himself for his country, he was not about to get out of the rocket without actually traveling in space. "I'm getting tired of this. Why don't you fix your little problem," he muttered through the private channel to the engineers, "and light the damned candle, cause I'm ready to go."[39]

Freedom 7 lifted off at 9:34 A.M., powered into space by an MR-7 Redstone rocket. Shepard flew to a height of 116 miles, got a short sample of weightlessness, and then splashed down off Grand Bahama Island just fifteen minutes later.

Cognizant of the need to sell space to the public, NASA paid close attention to how the flight would be perceived. The bigwigs decided that the entire flight would be televised, both to allow Americans to share in the excitement from start to finish, and to make a point about the more secretive Soviet space program. Powers told the press: "We have a calm, cool and collected astronaut." In fact, as Duff later admitted, "He didn't have any more idea that it was a calm, cool and collected astronaut than some guy in the press box making that same statement about a quarterback down on the field. It was totally subjective on Shorty's part. But he thought his job was to inject color into the program."[40] Powers also reported that Shepard had said that everything went "A-OK," an expression journalists loved and quickly incorporated into their repertoire of official space lingo. In fact, transcripts from the flight reveal that Shepard never said anything of the sort. He hated the expression.

The reception given to the flight surprised even Kennedy. Forty-five million Americans watched on live television, and 250,000 turned out to honor Shepard at a ticker tape parade in New York. Within days of the flight, a new school in Deerfield, Illinois, was named in his honor, and an unusual number of newborn babies were named Alan. At a Rose Garden ceremony, the astronaut was awarded the Distinguished Service Medal. The crowds who greeted him in Washington were bigger than those that had attended Kennedy's inauguration.

On May 8, Kennedy received a memo from Webb and McNamara that harmonized perfectly with the argument Johnson was pressing. Webb had earlier been reluctant to support a lunar mission because he feared that Kennedy might not be willing to fund such a venture to the extent necessary. Johnson, with help from Senator Robert Kerr, convinced Webb that money would not be a problem. Webb therefore felt sufficiently confident to tell Kennedy (in the memo compiled with McNamara) that "it is man, not merely machines in space that captures the imagination of the world. Dramatic achievements in space . . . symbolize the power and organizing capacity of a nation. It is for reasons such as these that major achievements in space contribute to national prestige."[41] After the Shepard flight, Kennedy fully agreed. Referring bitterly to the obstacles erected by the president's science adviser, Gilruth recounted: "When Kennedy saw how the American people loved that flight, it was all over as far as Weisner was concerned."[42]

The Shepard flight was the last piece in the puzzle. Kennedy now felt confident that the public would support the lunar mission Johnson had proposed. Around this time, Gilruth told Kennedy why a lunar mission made sense:

> I said, "Well, you've got to pick a job that's so difficult, that it's new, that they'll have to start from scratch." They just can't take their old rocket and put another gimmick on it and do something we can't do. It's got to be something that requires a great big rocket, like going to the Moon. Going to the Moon will take new rockets, new technology, and it you want to do that, I think our country could probably win because we'd both have to start from scratch.

The endeavor was also finite, the better to sustain public support. "I think that . . . Kennedy said we were going to the Moon not on the basis of what NASA had told him, but on the basis of how he felt as President he could bid for support," Webb reflected.[43]

Kennedy consulted key congressional leaders to mobilize bipartisan backing. After the immense enthusiasm shown for Shepard's flight, it took a brave legislator to refuse to board the lunar bandwagon. With all his ducks in a line, he then formally asked Congress to support a bold new initiative. On May 25, 1961, he told Congress: "Recognizing the head start obtained by the Soviets . . . and recognizing the likelihood that they will exploit this lead for some time to come in still more impressive successes, we nevertheless are required to make new efforts on our own. For while we cannot guarantee that we shall one day be first, we can guarantee that any failure to make this effort will make us last." Then came the stirring bit: "I believe that this nation should commit itself to achieving the goal, before this decade is out, of landing a man on the Moon and returning him safely to the Earth. No single space project in this period will be more impressive to mankind, or more important for the long-range exploration of space; and none will be so difficult or expensive to accomplish." He stressed that the mission would be a collective effort. "It will not be one man going to the Moon . . . it will be an entire nation. For all of us must work to put him there."[44]

Gilruth thought the decision owed a lot to Kennedy's youth. "He was a young man. He didn't have all the wisdom he would have had. If he'd been older, he probably never would have done it."[45] The national security adviser McGeorge Bundy thought the same thing. He remarked to Kennedy that the decision seemed like a "grandstand play" and one which only a supremely confident or supremely stupid leader would attempt. Kennedy replied: "You don't run for President in your forties unless you have a certain moxie."[46]

No matter how pretty the prestige, the cost of a lunar mission was obscenely huge. In 1961, the total federal budget was $94 billion. During his speech of May 25, Kennedy had asked Congress for an additional $531 million for fiscal year 1962, and a five-year commitment of between $7 and $9 billion in order to put a man on the Moon. If anything, those requests were misleadingly conservative. When Webb asked his team what it would cost to get to the Moon, the word came back: $10 billion. In order to be on the safe side, he applied what he called his "administrator's discount"—he took the estimate, doubled it, then doubled it again. He went to Congress and told legislators that the mission would cost just $20 billion, certainly no more than $40 billion. Overestimating the cost meant that NASA avoided invasive accounting by worried congressmen.[47] In his inaugural address, Kennedy had announced that a "new generation" of Americans was willing to "pay any price, bear any burden, meet any hardship . . . in order to assure the survival and the success of liberty."[48] Apparently, he wasn't kidding.

This commitment had to be made at a time when Americans had so far spent only fifteen minutes in space and when plans for going to the Moon were embarrassingly sketchy. But Kennedy was unrepentant. "If we are to go only half way, or reduce our sights in the face of difficulty," he cautioned, "in my judgment it would be better not to go at all." He claimed that "it is not a pleasure for any President of the United States . . . to come before the Congress and ask for new appropriations which place burdens on our people. I came to this conclusion with some reluctance. But in my judgment, this is a most serious time in the life of our country and in the life of freedom around the globe."[49]

Within Congress, opposition to Kennedy's plan arose primarily because of its huge cost. A significant number of conservative Republicans agreed with Eisenhower that an ambitious space program was an unaffordable luxury. They included Gerald Ford and Barry Goldwater. Their opposition is significant given the fact that both men were Cold War hardliners. They obviously did not believe that space was the best or most effective place to confront the Russians. In addition, conservative southern Democrats, like Richard Russell of Georgia, who were traditionally suspicious of massive federal programs, felt decidedly uneasy about Kennedy's big idea.

Opposition to the idea was outspoken, even influential, but not very numerous or strong. The ground was prepared by Johnson, a master at getting his way with Congress. Whenever he encountered opposition, he simply put his arm around the doubter, applied a gentle squeeze and asked: "Now, would you rather have us be a second-rate nation, or should we spend a little money?"[50] Opposition promptly melted. Kennedy's request for an immediate injection of new money for the space program was passed without much fight. But one has to question the nature of this consensus. Why were so many people so keen to go to a worthless rock in the sky? Part of the reason was the worship of technology. At the time, technology was synonymous with progress, therefore the process of going to the Moon was assumed to be a good thing—the mark of a dynamic nation and a process that would inevitably yield widespread benefit. In comparison, Eisenhower's fiscal prudence and belief in minimalist government seemed very old-fashioned.

Americans thought in terms of black and white. Going to the Moon would bring untold riches, staying on Earth would condemn them to poverty—of politics, economics, and of the soul. The benefits of a lunar landing could not, it was assumed, be had in any other way, nor did many people question whether those benefits were necessary or affordable. Few people listened to or accepted Eisenhower's argument that the United States was a great country, would remain a great country, and did not have to engage in silly displays of virility in order to demonstrate her greatness.

In 1961 Albert Wheelon and Sidney Graybeal, two intelligence analysts, likened the space issue to that of varsity football at America's premier universities. "Although both alumni and coach recognize that football has little to do with the true purpose of a college, the coach is under relentless pressure to win games because his team, in some intangible sense, stands for the entire college." Put in those terms, the whole crazy competition becomes more understandable. Every autumn, some of the greatest minds on Earth go potty if Harvard beats Yale in the annual gridiron grudge match. "It is much the same in the space race, a game which is similarly characterized by lively competition on the playing field and intense partisan interest among the spectators," argued Wheelon and Graybeal. "In a way that is neither rational nor desirable, our stature as a nation, our culture, our way of life and government are tending to be gauged by our skill in playing this game."[51]

One of Kennedy's favorite themes was the New Frontier. The concept had manifold domestic, Earth-bound manifestations, but was also perfectly suited to space, both as metaphor and as reality. The mission itself metaphorically suggested a dynamic, creative enterprise suitable to a modern society. But space itself was physically a new frontier, or at least Kennedy liked to portray it as such. It was, he claimed, the logical extension of the American West, a place where a new generation of intrepid American explorers would demonstrate their ingenuity and mettle. Adventures were believed to be essential in order to maintain a country's dynamism and to prove its virility. America had become a "rather settled society," according to the president, therefore it was important for the astronauts to demonstrate "that there are great frontiers still to be crossed."[52] Johnson agreed, claiming that the astronauts, like the pioneers, sought only to discover a better life and to secure their freedom.

In talks around the country, Webb reiterated this frontier myth. The "'wild and unperturbable' forces of the frontier, which show no mercy and no compassion, must be harnessed and utilized by the pioneer and in the process have the feedback effect of generating in the pioneer those qualities which have made for the American democratic system." In space, "an entire nation is developing technology which puts it, as an organized entity, very much in the same position as the pioneer was individually on the frontier."[53] In the space frontier, the Soviets were metaphoric Indians, aggressive competitors whose intent was malicious. Playing on this theme in his August 10, 1961, press conference, Kennedy told reporters, "We cannot possibly permit any country whose intentions toward us may be hostile to dominate Space."[54] One advantage of the frontier myth is that it led automatically to an assumption that settlers would follow the first space pioneers. Americans eagerly signed on to this preposterous idea. In the Senate, Clinton Anderson of New Mexico could hardly contain himself: "As surely as the ancient mariners ventured across unmapped seas, as certainly as explorers opened up our West, and men established bases at the barren and inhospitable poles, we are going to continue this bold journey through space."[55] Johnson said that "where the Moon is a major goal today, it will be tomorrow a mere whistle stop for the space traveler."[56]

The frontier metaphor had serious flaws. The American West, or Antarctica, or Mount Everest, were all big, but still finite. They could be conquered. The West could even be domesticated. The same could not be said for space, which is infinite. Orbiting the Earth in a tiny capsule could not by any means be seen as "conquering" space, no matter how often Kennedy presented it as such. The Moon might be conquered, but conquering it was a far cry from conquering space. The space frontier was in fact a sham—a confidence trick. In the great interstellar journey, landing on the Moon was hardly even a first step. The Moon as frontier was effective because people are familiar with it. They could see it, but at the same time it seemed far away and almost unattainable. To attempt a journey there, consequently, seemed significant. A lunar mission was a clever way of sidestepping the impossible infinity of space. Von Braun admitted as much when he told Congress:

> Everyone knows what the Moon is; everyone knows what this decade is; and everyone can understand an astronaut who returned safely to tell the story. An objective so clearly and simply defined enables us to translate the vague notion of conquering outer space into a hard-hitting industrial program that can be orderly planned, scheduled and priced out.[57]

But what then? The trick could be played only once. The rest of the frontier was overwhelmingly infinite. Most Americans could not even locate Mars in the sky.

The frontier myth shaped the way the mission was engaged. The goal was slave to the process, rather than the other way around. The process was devised in order to sustain the myth, not in order to achieve some useful purpose. In other words, the myth demanded that a frontier had to be explored by men, not by robots. Manned missions became inevitable, even though they were illogical. At some point in the race it was decided that space travel had to be an ego trip. Consequently, the limiting factor became that egotism. The decision that exploration had to be carried out by living, breathing human beings made missions much more complicated than they ever needed to be. Capsules had to be designed to keep human beings alive in all the threatening conditions of deep space. This placed severe limitations on what could be explored. Man can't land on Venus, where the surface temperature is 800 degrees Fahrenheit, but a robot can. Distances also became self-defeating. Insisting that the explorer must be a man implied that the mission had to be a round trip, which effectively ruled out any venture farther than Mars. Four days before Kennedy's speech, the NASA deputy administrator, Hugh Dryden, was asked about the practical benefits of putting a man on the Moon. "It certainly does not make sense to me," he replied.[58] After the speech, that kind of candor was seldom heard again.

As Johnson had outlined in his memo, the main reason to go to the Moon, or indeed to do anything in space, was prestige. Americans feared that Soviet space exploits would damage the reputation of the United States and cause countries around the world to go communist. "The Soviets have demonstrated how effective space exploration can be as a symbol of scientific progress and as an adjunct of foreign policy," Webb warned Kennedy on one occasion. "We cannot regain the prestige we have lost without improving our present inferior booster capability, and doing it before the Russians make a major break through into the multi-million pound thrust range."[59] Dryden, who did not see much scientific benefit in putting a man on the Moon, nevertheless accepted that the act of doing so would be indispensable as "a symbol." Journalists willingly signed on to the prestige argument. In *Newsweek*, Raymond Moley commented that space offered an opportunity for Americans to renew their faith in the American dream. "We need this renewal of faith even more than we need to reach the Moon."[60]

Not everyone agreed. Vannevar Bush called the man-in-space idea "a vastly overrated stunt." "There is nothing a man can do in space that can't be done better and more cheaply by instruments," he argued. As for the idea that nonaligned countries might be inclined to follow the superpower with the best record in space, he remarked: "I think a Guatemalan Indian is much more interested in getting a US drug to cure his sick child than he is in a US astronaut riding through space."[61] Glennan, the former NASA administrator, thought that Kennedy's challenge was "a very bad move." He warned that "we are entering into a competition which will be exceedingly costly and which will take up an increasingly large share of that small portion of the nation's budget which might be called controllable." As for the matter of America's reputation, he told Webb that "I cannot bring myself to believe that we will gain lasting 'prestige' by a shot we may make six to eight years from now. I don't think we should play the game according to the rules laid down by our adversary."[62]

The best rebuttal of the prestige argument, however, came from Eisenhower, who had struggled so bravely to resist it during his presidency. He thought that Kennedy, rattled by the Bay of Pigs fiasco, was plunging blindly into a commitment he would come to regret. "Why the great hurry to get to the moon and planets?" he asked. If prestige was so important, why not accentuate the important areas in which the United States was unquestionably superior to the Soviet Union? "Let us point to our industrial and agricultural productivity; why let the Communists dictate the terms of all the contests?" He reminded Americans that "in everything except the power of our booster rockets we are leading the world in scientific space exploration. From here on, I think we should proceed in an orderly, scientific way, building one accomplishment on another, rather than engaging in a mad effort to win a stunt race."[63] The whole venture, he decided, was "just nuts."[64] The nation, however, had stopped listening to Ike. In Congress, the Oklahoma Democrat Victor Wickersham countered that only a nut would "want . . . to live on Earth with a Russian man on the moon."[65]

Since not everyone accepted the importance of symbols, the space agency decided to play up the scientific potential of the lunar quest. An internal NASA document advised Webb, if asked about the value of a manned mission, to offer a stock reply: "I recently heard one well qualified scientist say that a geologist spending thirty minutes on the Moon would bring back more information than we would gain by placing 100 instrument packages on the Moon."[66] The statement was completely ludicrous, but revealing for being so. NASA clearly thought that it could blind the American people with science. NASA was probably right.

Occasionally, space agency officials could not stop themselves from telling the truth. In an article in *The Atlantic*, Homer Newell, director of NASA's Office of Space Sciences and Robert Jastrow, a NASA physicist, admitted that "while science plays an important role in lunar exploration, it was never intended to be the primary objective of that project." American efforts in space would be shaped "not by the measured patterns of scientific research, but by the urgencies of the response to the national challenge."[67] In other words, science was just a cloak used to hide a political demon.

Kennedy nevertheless stressed that science and education in the United States would be "enriched by new knowledge of our universe and environment, by new techniques of learning and mapping and observation, by new tools and computers for industry, medicine, the home as well as the school."[68] Providing backing vocals, Johnson heralded a "second industrial revolution" and repeatedly spoke of subsidiary benefits, or what is commonly called spin-off.[69] He maintained that Americans would reap "tangible benefits" in their daily lives from improvements in travel, communication, office equipment, and even "pots and pans." Every $1 spent on space, he said, would yield $2 worth of benefit. NASA trumped that by boasting in 1968 that "every dollar invested in the US space program 10 years ago is now returning four dollars."[70] Joining the chorus, Webb said that 3,200 space-related products would bring direct benefits to the American people, a figure for which he was, perhaps fortunately, not required to provide detailed evidence.[71] Nevertheless, included in NASA's spin-off list was a graphite substance used in the construction of tobacco pipes, a cigar-sorting machine, and a coating substance that speeded up hat blocking. "Hearing in a deaf person has been restored electronically via an operation and surgical implant of a tiny electronic device," NASA claimed. On release of that report, the company developing the device was forced to admit that it had been tried only once and was unsuccessful. "We don't even like to talk about it for fear of prematurely raising hopes of many deaf persons," an official remarked.[72] In 1963, Webb claimed that the lunar mission had resulted in the development of the weather satellite, which "is said to be saving us an estimated billion dollars a year in property losses."[73] While no one would question the value of weather satellites, Webb seems to have forgotten that the first one went up before the decision to go to the Moon was taken.

Before long, NASA claimed that Teflon, the miracle product that kept eggs from sticking, was a "major example of space spinoff."[74] In fact, the substance had been developed by DuPont in 1938 and was first used extensively in the Manhattan Project. A similar false claim was made with respect to Velcro. On another occasion, it was claimed that "time-saving pots and pans are now coated with a plastics material developed to protect spacecraft from the extreme heat of launch and re-entry."[75] In fact, the technology traveled the other way: the Corning company sold coating materials already present in American kitchens to NASA. The agency also had a tendency to look for complex solutions when simple ones would have sufficed. Vast resources were spent in developing a pen that would work in zero gravity. That invention was then celebrated by NASA PR and marketed to space enthusiasts. The Russians, on the other hand, made do with pencils.

Over the years, quite a few self-congratulatory pamphlets and books produced by NASA detailed all the wonderful products the agency gifted to mankind. But this is a rather roundabout way to produce useful consumer goods. A lover of irony, George Washington University professor Amitai Etzioni (later at Columbia) admitted that "if you burn . . . 25 billion dollars you're gonna get SOME light."[76] "If you want the by-product," Stanford physicist Wolfgang Panofsky quipped, "you should develop the by-product."[77] Other critics pointed out that new technologies often merely addressed problems that NASA itself created by putting humans in space. Only an astronaut needs to know how to make urine potable.

"We've disproved the critics who said that going to the Moon is not going to help solve problems here on Earth," Richard Lesher, NASA's assistant administrator for the Office of Technology Utilization boasted. "A spacecraft is a small planet in which you have water and air pollution, physiological, engineering and all sorts of problems. So, whatever we learn in those fields will carry through to the solution of problems on earth."[78] Rarely have public agencies invested so much money and effort into proving their virtue. During its first decade of existence, NASA funded social scientists to the tune of $35 million in an attempt to quantify the social and economic benefits of the space program.[79]

Eisenhower had insisted that the American space effort was entirely benign and not connected in any way with military goals. Kennedy did likewise, but he still rather cleverly suggested that failure to perform impressively in outer space would weaken American security. Johnson chimed in, arguing that "Americans do not intend to live in a world which goes to bed by the light of a communist Moon." From NASA, D. Brainerd Holmes, director of Manned Space Flight, insisted that the Moon mission would "advance the welfare and security of the nation." Within Congress, there were plenty of firebrands who thought that the connection between space and national defense was unambiguous. "I can't see how there can possibly be any doubt that there is a military mission in space," argued Texas Democrat Olin Teague, a member of the House Committee on Manned Space Flight. John Stennis, a Democrat from Mississippi, "very strongly" agreed that "there is a great military value to this space program."[80] This sort of attitude was useful to Kennedy for the very simple reason that those Americans who might otherwise have been unimpressed with esoteric notions of space adventure or the need to explore were swayed by an argument based on national defense. Editorials in small provincial papers like the *Massapequa Post* of New York indicate that there were plenty of true believers in the military value of space. "Doesn't history show that overindulgent, complacent civilizations are invariably swept away by vigorous, warlike peoples?" the editor asked. "Let's not civilize ourselves right out of existence. In the abstract, we agree that schoolhouses are more important than space probes, but let's make sure that we retain the freedom to control the education in them."[81]

Clearly, there was something for everyone out in space. Kennedy's justification for the program was like a medieval castle with concentric walls of defense. Some critics could scale one or two walls, but few could scale all of them and expose the sham at the center. The first wall was labeled science, the second defense, the third frontier adventure, and the fourth prestige. Critics of the space program exhausted themselves trying to knock down the bastion Kennedy had built.

Since the American program was born of desperation and insecurity, and was intended as an answer to the Cold War problem of prestige, opposition to it was seen as anti-American. How could anyone oppose an effort so sublime, and a program so innocently noble? Only the most confident, independent, and secure congressman was prepared to shout, on the floor of the House or Senate, that the emperor had no clothes. The weakness of the opposition can be explained in part by the fact that it came from two completely divergent political directions. Liberal progressives like J. William Fulbright, Ernest Gruening, Maureen

Neuberger, and Joseph Clark repeatedly argued that the billions wasted on the Moon mission should be spent instead on schools or hospitals back on Earth. Fulbright called the program a "gaudy sideshow" and argued that "not a single nation" had gone over to the Soviet side because of Russian exploits in space.[82]

Liberals were, however, never likely to find consensus with politicians like Goldwater and Howard Cannon, who objected that, instead of going peacefully to the Moon, the United States should invest billions in putting weapons into space. Representative James Weaver insisted that the Kennedy administration should embark on the space mission with national security as its overriding concern. "Congress and the public can no longer tolerate public relations gimmicks and doubletalk concerning the space program and our space gap when national security is threatened," he said. Another hardline congressman was Donald Rumsfeld, who insisted that "this country should direct itself toward inner space and not place our top priority in the direction of the Moon."[83] "Only the vain and incurably sentimental among us will lose sleep simply because foreign people are not as impressed by our strength as they ought to be," Goldwater remarked with typical acuity.[84] As he pointed out, showy attempts to prove one's worth merely invited ridicule. A quiet nation at peace with itself was more likely to gain respect or, as Teddy Roosevelt once said: "Talk softly and carry a big stick."

By stating a goal, Kennedy created an expectation. Americans had been told that going to the Moon was a good thing, so they naturally concluded that the task would be pursued to completion. For Kennedy, the idea at first seemed sensible—proof of his dynamism and the solution to so many of his problems. It would put Americans to work, unite them behind a common goal, and allow them to show the Russians just what they were worth. But the cost was astronomical. At the time, the budget for the War on Poverty was $1.8 billion. Elementary and secondary education were getting just $2 billion in federal money. The real worry for Kennedy was that he would not be able to go to Congress for further huge appropriations for programs much closer to his heart. But he could not kill the program because he had sold it by arguing that it was essential to the security of the United States. The space program was ring-fenced by Cold War fears.

The *Washington Post* reported in April 1963 that a visitor to the White House challenged Kennedy on whether it really mattered if the United

States went to the Moon. In a brief moment of candor, the president apparently replied, "Don't you think I would rather spend these billions on programs here at home, such as health and education and welfare? But in this matter we have no choice. The Nation's prestige is too heavily involved."[85] According to Weisner, Kennedy saw the Moon as a solution to an earthly dilemma. "We were paying a price—internationally, politically—and that was the issue that the president was dealing with, not was it time to go to the Moon or not, but how to get yourself out of this." He continued: "We talked a lot about do we *have* to do this." Kennedy replied: "Well, it's your fault. If you had a scientific spectacular on this Earth that would be more useful—say desalting the ocean—or something that is just as dramatic and convincing as space, then we would do it." Weisner was always of the opinion that "if Kennedy could have opted out of a big space program without hurting the country in his judgement, he would have. . . . I think he became convinced that space was the symbol of the twentieth century. It was a decision he made cold bloodedly. He thought it was right for the country."[86]

Thanks to Kennedy, Americans were on their way to the Moon. But the reasons they were going were not those of dreamers like Tsiolkovsky. Going to the Moon had nothing to do with exploring the cosmos or understanding the origins of the universe. The decision was based not on science, but on cold hard politics. McNamara needed to save the aerospace industry. Johnson wanted to restore American prestige. Congress worried about losing influence in the Third World. Senators wanted fat contracts for their states. Kennedy needed to rescue his image. Everyone wanted to beat the Russians. America would go to the Moon for all the wrong reasons. If good reasons ever existed, they had been buried under a pile of politics.

The shallow nature of the contest can be judged by one politician's reaction to a rumor circulating in July 1961. Referring to the fact that the Moon had become a visible "propaganda symbol," Fulton revealed that the Russians might be planning to fill a rocket nose cone with red powder and crash it into the Moon. If this happened, he advised, the Americans should respond with a "blue project"—crashing a nose cone filled with blue powder. At the time it seemed entirely sensible to redecorate the moon in red, white, and blue.[87]

9

The Sleep of Reason Produces Monsters

When James Webb took over NASA, around 6,000 people worked for the agency. When he left in 1968, his payroll numbered nearer 60,000. But that's no indication of the real growth of the space industry, since seven out of every eight workers involved in Apollo were employed by private contractors.

Now let's suppose that every day, each of those 500,000 workers spent a dime on a candy bar from a vending machine at the plant. Okay, some might have been on a diet, but others perhaps had a Babe Ruth and a Butterfinger on the same day, so dieters and bingers probably canceled each other out. That meant a gross return of $50,000 per day from candy on the Apollo project alone. Now that would have been a good franchise to have.

The man who had it was Bobby Baker, a Washington wheeler-dealer who in 1967 was found guilty of seven counts of theft, fraud, and income tax evasion. Among Baker's friends were some of the shadiest characters of the 1960s underworld. He not only owned the NASA candy machine franchise, he also had a near monopoly of the cigarette and soft drink machines—not just in the Apollo plants, but throughout the defense industry. (The machines themselves were made at a factory owned by the mobster Sam Giancana.) The man who helped Baker get that franchise was his friend and mentor, Senator Robert Kerr of Oklahoma. Kerr made his money through Kerr-McGhee Oil Industries, one-time employer of Webb. He was also a great friend and supporter of Johnson, who had consulted Kerr while compiling his paper in support of going to the Moon. According to the journalist Milton Viorst, "Kerr was a self-made millionaire who freely and publicly expressed the conviction that any man in the Senate who didn't use his position to make money was a sucker."[1]

And you thought Apollo was a story about heroes.

. . .

Kennedy had set a goal of getting to the Moon. Lots of intelligent scientists and engineers were thinking about the idea, but none had a firm sense of how to do it. The simple way was to aim a rocket directly at the Moon, slow it down when it approached the lunar surface, and then land. But that meant that the vehicle which took astronauts to the Moon, and took them back, would actually have to land on the surface. Getting that kind of weight off the surface would require a pretty big rocket, about 80 feet high. That would be like landing an eight-story building on its end on the Moon. Okay, suppose that's possible. How then does the astronaut get down? That might be easy, since the Moon has some gravity which would pull a person to the surface, as long as he doesn't fall too fast. But how does the poor guy get back into the capsule for the ride home? Forget all that, since the real problem would be getting a contraption that size off the Earth in the first place. It would have required a rocket bigger than the Americans ever built, something like a seven-stage behemoth weighing about 14 million pounds.

Two solutions were proposed: Earth Orbit Rendezvous (EOR) and Lunar Orbit Rendezvous (LOR). Both involved a smaller craft separating from the main capsule and descending to the lunar surface, thus requiring much less thrust to leave the Moon. The choice came down to whether the lunar module would separate from the command module while in Earth orbit or whether the joined craft would travel together to the Moon and the separation would take place in lunar orbit. Both had their advantages, both their disadvantages. Each was incredibly complex, and for a long time it didn't look like the complexities could possibly be solved in time to meet Kennedy's deadline.

John Cord and Leonard Seale, engineers at Bell Aerosystems, proposed another idea that eventually became known as the "poor slob" plan, though not by them. Maybe it was just a joke, but, given the urgency of Kennedy's goal, it was more likely offered in complete seriousness. It went like this: getting to the Moon was easy, getting off was hard. So, we send some (unmarried) guy to the Moon with lots of food and a good supply of really long books. We leave him there, occasionally sending him more food and more books via resupply ships. In the meantime, we work on the problem of how to get the poor slob back to Earth. We win the race to the Moon, and the poor slob gets a lot of reading done. Who knows? We might someday even get him home.*

That plan, however, seemed just a bit too uncivilized for American tastes, rather like something the Russians might try. So, minds concentrated on the other two ideas. EOR was von Braun's baby. LOR was proposed by a man named John Houbolt. Critics said that Houbolt's idea had "a 50 percent chance of getting a man to the Moon, and a 1 percent chance of getting him back."[2] Carrying out a rendezvous in lunar orbit seemed far too scary. But the thing in its favor was that it kept the weight of what had to be landed on the Moon down to an absolute minimum. Eventually the NASA establishment finally came round to LOR. The deal was done when, on June 7, 1962, von Braun suddenly switched sides. The motivating factor was time, as the aerospace engineer Jesco von Puttkamer explained:

> We in Huntsville had proposed Earth Orbit Rendezvous. . . . This would have required tanker flights and a larger ship, which would have been assembled from modules in orbit around Earth, something which comes pretty close to building a space station. We had designed a tanker ship, and there were already concepts of a space shuttle. We would have had to develop refuelling techniques in Earth orbit, but we didn't see any problems in doing that. However, some other people in Houston decided to go to LOR. . . . I think one of the arguments was that the LOR ship would have taken less time to build—eight years they projected. Kennedy, of course, wanted us to land on the moon during his second term.[3]

NASA opted for the quick fix, even though, as deputy administrator Hans Mark admitted, "Apollo was essentially a dead-end from a technical viewpoint."[4] LOR was a single-purpose venture—it would get an American to the Moon (and back), but didn't lead to anything. Von Puttkamer deeply regretted the decision:

> The reason some of us wanted EOR was not just to go to the Moon but to have something afterwards: orbital operations, a space station, a

* According to William Burrows (*This New Ocean*, p. 370), a cosmonaut named Mikhail Bordeyev, around the time when it became clear that the USSR would lose the race to the Moon, volunteered to be a poor slob on a mission to Mars.

springboard. LOR was a one-shot deal, very limited, very inflexible. But if you developed an Earth-orbiting space station first, you would have had the flexibility to keep ships going to the Moon, to land there and stay there, or to go to Mars, or just to exploit the capability of Earth-orbital operations. After the close-down of Apollo, we began to pay the price.[5]

Doing something meaningful in space was never really the point. The goal was to get to the Moon, quickly, and then forget the whole thing. The other idea made scientific sense, but science wasn't the issue.

After Alan Shepard's flight, the Mercury capsule was "improved" slightly with the addition of a large window (essential to call the pilot a pilot) and an explosive hatch to allow for quick escape. The next astronaut was Gus Grissom. Again, a short fifteen-minute suborbital mission was planned. The flight itself went well, but, on splashdown, the explosive bolts fired and the capsule began to sink. Grissom subsequently claimed he did not touch the trigger mechanism for the hatch, but a malfunction seems unlikely. He got himself out of the capsule, only to find that his spacesuit quickly filled with water, threatening to drown him before the recovery helicopter could arrive. It didn't help that his pockets and boots were stuffed with coins—Space Dimes that he wanted to distribute as souvenirs after the flight. The space program came within seconds of producing its first fatality.

Just over two weeks after Grissom's short flight, the Russians again blew a big raspberry at the United States. On August 6, 1961, *Vostok 2*, with Gherman Titov aboard, spent an entire day in space, completing seventeen orbits. The press again started to get uneasy about the huge gap between American and Soviet accomplishments. The euphoria that followed Shepard's flight and Kennedy's speech had dissipated. On October 11, 1961, at a White House press conference, a reporter asked Kennedy whether the nation had reacted positively to his landmark speech. "Until we have a man on the Moon, none of us will be satisfied," he replied, "But . . . as I have said before, we started far behind, and we are going to have to wait and see whether we catch up. But I would say that I will continue to be dissatisfied until the goal is reached, and I hope everyone working on the program shares the same view."[6] No one was quite sure what that meant.

In 1961, the United States shot forty-one rockets into space. Of those, twenty-nine were successful. The Soviet Union, which was running short of R-7 rockets and Vostok capsules, had five successful launches and four failures.[7] By that statistic alone, it would appear that the United States was already ahead in the space race and likely to stay there if only because of its economic ability to continue producing hardware. But none of this mattered to the American people, nor to the world. What mattered was that the Russians had put two men in orbit on separate occasions and, in contrast, the United States had managed only two puny suborbital flights. Had the real picture been available in 1961 and had logical minds been around to analyze that picture, it would have been patently clear that the United States held the advantage. Instead, a great deal of envy and fear remained to be expended.

After Titov's flight, the pressure was on for more ambitious missions, but NASA remained cautious. In order to emulate the Russians and put a man in orbit, the agency would have to abandon the Redstone rocket in favor of the more powerful, but less reliable, Atlas. In mid-September 1961 an unmanned flight proved successful, and it was followed, on November 29, by another Mercury capsule piloted by a small primate named Enos. He was a clever chimp, and, like Ham, well trained. He knew all about how to avoid getting zapped with those shocks to his toes. But he had one unfortunate habit—he loved to masturbate. The medical technicians decided that this was not sufficiently serious to disqualify such a talented primate from becoming the first American to orbit the Earth. In any case, he'd be wearing a urinary catheter in space, which, it was thought, would curb his desires.

The launch went smoothly, but then the environmental control system went awry, sending the temperature inside the capsule above 100 degrees Fahrenheit. Meanwhile, the chimp was busily performing his reaction tests in order to avoid getting shocked. Unfortunately, the little machine malfunctioned with the result that he got shocked even when he pulled the correct lever. He naturally got angry and started tearing the equipment apart. Whether out of rage or sexual frustration, he yanked the catheter off his penis, which must have been painful. Worried NASA controllers decided to shorten the flight to just two orbits. A very hot and extremely angry chimp returned to Earth after a little over three hours in space. The next day, at his postflight press conference, he horrified his NASA handlers when he ripped off his diaper and started to fondle himself. Journalists would forever remember Enos the Penis.[8] NASA, on the other hand, was probably delighted that the era of chimps in space was drawing to a close. John Glenn could be trusted not to play with himself.

Chimp was supposed to give way to man, but Glenn's launch, originally intended as a Christmas present to the American people, was postponed nine times over a total of eighty-two days. Meanwhile, the American public was growing increasingly impatient. Kennedy had promised a race, yet the Americans didn't seem to be running. On February 7, 1962, the president was asked: "Have we changed our timetable for landing a man on the Moon?" Kennedy again reminded his listeners that the United States had started the race far behind the Russians and could not therefore expect to overtake them quickly. "We, however, are going to proceed," he stressed. "To the best of my ability, the time schedule, at least our hope, has not been changed by the recent setbacks."[9]

Glenn's *Friendship 7* finally lifted off on February 20. In Cocoa, time stood still as people lined the beaches and roads to watch the liftoff. One hundred million people watched on television.* ABC, NBC, and CBS averaged twenty-nine hours of coverage pertaining to the mission. The flight itself seemed to go well, but then, out of the blue, came a hazard warning detected by controllers on the ground. It indicated that the heat shield was not properly locked in place, when logic suggested otherwise. Mission Control was faced with the possibility that the heat shield might fall off during reentry, turning the capsule into a furnace. At the very least, even if the shield remained attached, extreme hot spots might develop, burning a hole through the capsule and frying the astronaut or suffocating him in noxious gases. The engineers felt that the warning had to be taken seriously. They decided that the six retro-rockets, which were attached to the heat shield and would ordinarily have been jettisoned prior to descent, would be kept in place, in the hope that the metal straps that fastened them would also hold the shield in place. What resulted was a spectacularly fiery ride back to Earth, as the retro-rockets, and the metal straps, burned away. A subsequent investigation revealed that there was no problem with the heat shield. The switch controlling the warning signal was faulty.

The crisis worked to NASA's advantage. Playing on the man versus machine theme so prevalent in space reportage, a *Newsweek* jour-

---

* In Atlantic City, New Jersey, Sister Mary Seinna appealed to the public to donate 72,000 Green Stamps so that her school, St. Nicholas, could obtain a television to allow students to watch Glenn. She had all the stamps she needed, and more, within four days.

nalist reminded readers that machinery had faltered, "never Glenn." A *New York Times* reporter likewise remarked that the mission had demonstrated that "we need not be ruled by machines."[10] Glenn himself quipped: "Now we can get rid of some of that automatic equipment and let man take over."[11] All of NASA's work in promoting the frontier message had worked perfectly.

There followed a coronation, the moment Americans had been waiting for since they'd first met Glenn. The Dave Brubeck Quartet released the album *Time in Outer Space,* dedicated to Glenn; while Sam "Lightning" Hopkins sang "Happy Blues for John Glenn," and Roy West crooned "The Ballad of John Glenn." Rio de Francisco wrote a song called "Spaceman Twist" which, after the flight, he released as "The John Glenn Twist."

> *Let's do the John Glenn Twist! Yeah! Oh!*
> *Round and round and a-round*
> *Three times around the world he goes,*
> *Up in space, Orbitin' in space*
> *And the whole wide world knows. Knows.*[12]

Hundreds of baby boys were given the name "John Glenn." Mr. and Mrs. Mario Garnica of Dallas had planned to name their son Robert Kennedy, but could not resist the lure of the astronaut. In Kazakhstan, Mr. and Mrs. Aleksandr Revkov named their triplets Yuri, Gherman, and John.[13] Perhaps the strangest commemoration of this sort, however, came when Mr. and Mrs. H. Roy Hill of Ogden, Utah, named their baby son Lamar Orbit.[14]

Like Shepard, Glenn went to Washington for a medal and a parade. A quarter of a million people lined Pennsylvania Avenue to catch a glimpse of his motorcade. He then addressed a joint session of Congress, a privilege usually reserved for world leaders. "I still get a hard to define feeling inside when the flag goes by—I know you do too," he told legislators.[15] They swooned. When Glenn addressed the Senate Aeronautics and Space Sciences Committee, the atmosphere, according to the *New York Times*, "took on the aura of a revival meeting in a tent in Oklahoma." Glenn told the committee that his was not a "fire-engine type religion" to be called upon only in a crisis, but instead part of his daily life. In a direct rejoinder to Gagarin, he insisted that God was in space just as on Earth. "I don't know the nature of God any more than anyone else," he said. But, "He will be wherever we go." Senator Kerr, the committee chairman, congratulated Glenn "on his spirit of reverence and his faith in God."[16]

Later came a special performance on Broadway of the hit *How to Succeed in Business without Really Trying*. Glenn and his family were late, but the performance was held up for them. They arrived to a standing ovation and found that the script had been hastily rewritten in order to include references to the flight. NASA, cognizant of a PR opportunity, then sent *Friendship 7* on a world tour. People in seventeen countries got a chance to see the little capsule in which Glenn circled the Earth. After the tour, the capsule took a brief stop at the Seattle World's Fair. The fair was a dramatic demonstration of how America had changed since Brussels four years earlier. Gone were the televisions, burgers, and ice cream. In came a Space Needle, a monorail, and Mercury. America was suddenly modern again, enlivened by the race with the Russians.

On March 7, 1963, Sara Bartholomae of Los Angeles announced plans to build a $1 million Mercury Chapel in honor of John Glenn atop a knoll overlooking Brea Canyon. She planned to finance the construction from the proceeds of her $4.5 million divorce settlement from tycoon William Bartholomae.[17] So great was the public reaction to Glenn's flight that Kennedy decided the astronaut was more valuable to the space program on the ground than he could ever be atop a rocket. He ordered that Glenn should not be given any subsequent missions, exactly the same decision Khrushchev made with regard to Gagarin.

The day after Glenn's flight, Kennedy told a press conference about a phone call he had received from Khrushchev congratulating the Americans on their success. The Soviet premier had gone on to suggest that "it would be beneficial to the advance of science if our countries could work together in the exploration of space." Kennedy announced that he would be replying positively to the message. "I regard it as most encouraging. . . . We believe that when men reach beyond this planet they should leave their national differences behind them. All men will benefit, if we can invoke the wonders of science instead of its terrors." When asked whether Khrushchev had made any concrete suggestions about areas of cooperation, Kennedy admitted that he had not. "But we, I might say now, have more chips on the table than we did some time ago, so perhaps the prospects are improving."[18] The reporters then pressed the point. Would cooperation with the Soviets stretch to a joint United States/Soviet effort to get to the Moon? On this point, Kennedy refused to be drawn. But he did stress that "we are spending billions of dollars in Space, and if it is possible to assure that Space is peaceful, and that it can be used for the benefit of everyone, then the United States must respond to any opportunity we have to insure that it's peaceful."[19]

The idea was not entirely new. A hint had come during Kennedy's inaugural address and had been reiterated on January 30, 1961, during his first State of the Union speech, when he revealed that his administration intended "to explore promptly all possible areas of cooperation with the Soviet Union." Cooperation, he maintained, made sense because "today this country is ahead in the science and technology of space, while the Soviet Union is ahead in the capacity to lift large vehicles into orbit. Both nations would help themselves as well as other nations by removing these endeavors from the bitter and wasteful competition of the Cold War."[20]

Kennedy, it seems, might have wanted to offer the Soviets the opportunity to call off the race in order to avoid a situation that threatened to cripple both countries financially. Rather than throw huge amounts of money at the Moon, Kennedy thought that it might be wiser to do a deal with the Russians and remove the importance of space from Cold War calculations. The idea would periodically resurface during Kennedy's remaining time in the White House. On March 23, 1962, he told a crowd gathered at the University of California that

> a cooperative Soviet-American effort in space science and exploration would emphasize the interests that must unite us, rather than those that always divide us. It offers us an area in which the stale and sterile dogmas of the cold war could be literally left a quarter of a million miles behind. And it would remind us on both sides that knowledge, not hate, is the passkey to the future.[21]

On September 23, 1963, in a speech before the United Nations, Kennedy even went so far as to propose a joint mission to the Moon.

> Why . . . should man's first flight to the moon be a matter of national competition? Why should the United States and the Soviet Union, in preparing for such expeditions, become involved in immense duplications of research, construction, and expenditure? Surely we should explore whether the scientists and astronauts of our two countries—indeed of all the world—cannot work together in the conquest of space,

sending someday in this decade to the moon not the representatives of
a single nation, but the representatives of all of our countries.[22]

Cooperation not only offered a way out of the expensive lunar commitment, it also presented the tantalizing possibility of a foreign policy
breakthrough so big that it would put the space race permanently into
the shade. Détente with the Soviet Union would allow Kennedy to be
forgiven for backtracking on his lunar pledge.

In truth, however, the idea was never more than pie in the sky.
Without the element of competition, there was no reason to go to the
Moon at all. The existence of a high-profile, supposedly important race
with the Soviets was the best reason NASA could give for its ever-in-
creasing demands for cash. The same held true for the Russians. As
Leonid Sedov, chairman of the Committee of Interplanetary Communications of the Soviet Academy of Sciences, remarked at the time, "If we
really cooperated . . . neither country would have a program because
the necessary large support in money and manpower was only because
of the competitive element and for political reasons."[23] For NASA to
back cooperation was essentially to acquiesce in its own emasculation.[*]

Leaving aside the fact that both the Russian and American space
programs were deeply entangled with secretive military projects (and
that is a big matter to leave aside), there was one huge problem with cooperation, namely that so much of what the Soviets had achieved was
based on bluff. They were doing rather well with their big rocket which
threw a variety of large objects into orbit. Huge propaganda gains had
been made as a result of this simple feat. The world had concluded that
the Soviets were masters of technology. So far, they hadn't put a foot
wrong in this bloodless battle for world prestige. Cooperating with the
Americans would throw all that out the window. They would have to
open up their space program to American eyes, and it would become
painfully obvious how limited Soviet capabilities actually were. Their
space program was like a great Potemkin village shot into the sky. They
knew how to build big boosters, but little else. In all the sophisticated
techniques needed to put a man on the Moon, the United States was already far ahead, even if this lead could not yet be demonstrated to the
world.

* The Senate did not much like the idea, either. In response to Kennedy's talk of cooperation, senators passed a rider to the 1964 budget that forbade any joint ventures without congressional consent.

An even worse situation faced the Americans. If they were to co-operate, the world would assume that they were the junior partners in the enterprise and that any subsequent achievements were Russian in-spired. To argue otherwise would seem churlish and would inevitably jeopardize the agreement to cooperate. In other words, neither side would ever be able to find a way of joining together while still saving face. The pointless race would therefore have to go on, since it was not the Moon that was important, but prestige.

The next astronaut in the queue was Deke Slayton, but a heart anomaly sidelined him. In his place stepped Scott Carpenter, whose *Aurora 7* lifted off on May 24, 1962. The mission included some experiments in-volving the behavior of liquids in space, designed in part to give the game of outer-space basketball some legitimacy by attaching a mantle of scientific inquiry.

A failure in the capsule's automatic control system allowed Car-penter to act like a pilot in a way that Shepard, Grissom and Glenn had been denied. While this malfunction might have underlined the impor-tance of having a man to step in when computers failed, in fact it re-vealed how human beings do not always perform as desired. His ma-neuvering consumed fuel at a faster than anticipated rate, seriously en-dangering the mission. Whether preoccupied or confused, Carpenter failed to fire the retro-rockets at the precise moment for reentry, which meant that he missed his target for splashdown by some 250 miles. NASA knew where he was, but it took some time for rescue ships to ar-rive. For nearly an hour, Americans chewed their fingernails as they waited for the tiny capsule to be located. The Soviets even offered their help, which the suspicious Americans, taking it as a publicity stunt, de-clined. Carpenter was eventually found sitting in a rubber life raft teth-ered to the capsule, which was moments from sinking. He later an-noyed NASA by telling a press conference "I didn't know where I was, and they didn't either."[24] He was never allowed to fly again.

In comparison to Carpenter's flight, the Soviet's next effort looked magnificent. Sergei Korolev, ever the showman, took two of his rockets and fired them on consecutive days. *Vostok 3*, carrying Andrian Niko-layev, lifted off on August 11, following by *Vostok 4*, piloted by Pavel Popovich. They assumed the same trajectory, but for the fact that Popovich was slightly higher in orbit. This had the effect of eventually bringing them to within 5 kilometers of each other. To those who did not

understand orbital motion, it appeared that the two spacecraft had executed the world's first rendezvous, an essential prerequisite to building a space station or getting to the Moon. In fact, they'd done nothing
of the sort. No maneuvering had taken place; physics and geometry had
simply brought the two capsules into close proximity and then, immediately afterward, took them away from each other. To complete the
trick, the Vostoks came down within six minutes of each other and just
200 kilometers apart.

The Soviets never actually claimed that they had carried out a rendezvous; they allowed journalists to draw that conclusion. A British
newspaper called it "TWO UPMANSHIP." Edward Teller, the professional alarmist, voiced the opinion that "there is no doubt that the best
scientists as of this moment are not in the U.S., but in Moscow." The
highly respected British astronomer Sir Bernard Lovell decided that, because the Russians were so far ahead in space, "the possibility of America catching up in the next decade is remote."[25] It was all a trick, but
NASA could not say so, for that would have seemed like sour grapes.

NASA was in the doldrums. The money to fund Kennedy's wild
ambitions was not yet flowing, and Congress seemed to be losing enthusiasm. While the president would have loved to find a way out of
the trap he had set, he had to appear the great champion of space exploration. On September 12, 1962, he gave a speech at Rice University
in Houston, not far from the new Manned Spacecraft Center. Speaking
in the football stadium, he sounded like a coach trying to inspire a demoralized team:

> The exploration of space will go ahead, whether we join in it or not,
> and it is one of the great adventures of all time, and no nation which
> expects to be the leader of other nations can expect to stay behind in
> the race for space. . . . This generation does not intend to founder in the
> backwash of the coming age of space. We mean to be a part of it—we
> mean to lead it. . . . Whether it will become a force for good or ill de-
> pends on man, and only if the United States occupies a position of pre-
> eminence can we help decide whether this new ocean will be a sea of
> peace or a new terrifying theater of war.

Kennedy was trying to create a broad constituency by drawing in those
people motivated by fear of the Soviets, but also those driven by something more sublime.

But why, some say, the Moon? Why choose this as our goal? And they may well ask why climb the highest mountain? Why, 35 years ago, fly the Atlantic? . . .

We choose to go to the Moon. We choose to go to the Moon in this decade and do the other things, not because they are easy, but because they are hard, because that goal will serve to organize and measure the best of our energies and skills, because that challenge is one that we are willing to accept, one we are unwilling to postpone, and one which we intend to win.

Kennedy implied that he had no regrets about the bold challenge made a year before. It would be seen as "among the most important decisions that will be made during my incumbency in the office of the Presidency."

There was, of course, the matter of cost. "This year's space budget is three times what it was in January 1961, and it is greater than the space budget of the previous eight years combined," the president admitted. "That budget now stands at $5,400 million a year—a staggering sum, though somewhat less than we pay for cigarettes and cigars every year." Space expenditure, he predicted, would soon rise from 40 cents per person per week to more than 50 cents. "I realize that this is in some measure an act of faith and vision, for we do not now know what benefits await us." In trying to justify the investment, Kennedy ended with reference to mountain climbing, another dangerous pursuit which, some would argue, was rather pointless.

Many years ago the great British explorer George Mallory, who was to die on Mount Everest, was asked why did he want to climb it. He said, "Because it is there."

Well, space is there, and we're going to climb it, and the moon and the planets are there, and new hopes for knowledge and peace are there. And, therefore, as we set sail we ask God's blessing on the most hazardous and dangerous and greatest adventure on which man has ever embarked.[26]

The United States would spend a lot of money on a great adventure. She would encounter danger. She might sacrifice lives. But it would all be worthwhile. What would it bring? Kennedy wasn't sure. But it had to be done. It had to be done because space was there.

In the following month came some good news. A mission virtually identical to Carpenter's was undertaken by Walter Schirra in *Sigma 7.* He managed to complete a flawless flight and a splashdown precisely on target. Around this time, Kennedy, interested in saving money, asked James Webb to explore the possibility of accelerating Apollo with a view to a lunar landing in 1966, even if that meant cutting back on all other space programs. Webb replied that the idea was unwise and potentially dangerous. He had other objections, which he hesitated to voice explicitly. As he later explained, the public's goal was not NASA's goal: "The lunar project for us was little more than a realistic requirement for space competence. . . . We realized there must be a continuing advance of scientific knowledge . . . and when the politicians, including the President, tended to say, 'Well, gee, we've got a tight budget, just concentrate on getting this Moon thing,' we always said, 'No, our objective must be broader.'"[27] Webb would struggle with this problem throughout his career at NASA. On one occasion, he moaned to Jerome Weisner: "I've fought tooth and toenail to avoid this implication that we are just a one purpose agency for going to the Moon."[28] He understood that the public and the president thought in terms of a single goal, but he also understood that such a goal was like fixing a date for the demise of NASA.

On November 21, an important meeting on space policy took place at the White House. Present were Webb, Weisner, Lyndon Johnson, Robert Seamans, David Bell, Hugh Dryden, and a number of other individuals from government and the space program. In the course of the meeting, the president asked if going to the Moon was NASA's top priority. Webb replied:

> No, sir, it is not. I think it is one of the top priority programs but I think it is very important to recognize here that as you have found what you could do with the rocket, as you have found all you could get out of it beyond your investment field and into space and make measurements several scientific disciplines that are very powerful begin to converge.

Kennedy rudely interrupted Webb while the latter was in full flow. The president was not pleased to hear that going to the Moon was not NASA's overriding priority. "This is important for political reasons, international political reasons," Kennedy insisted. "And this is, whether we like it or not, in a sense a race. Being second to the Moon is nice, but it's—it's like being second anytime." He gave Webb an earful:

Everything that we do should be tied into getting on to the Moon ahead of the Russians. We ought to get it really clear that the policy ought to be that this is the top priority program of the agency and one . . . of the top priorities of the United States government. . . . Otherwise we shouldn't be spending this kind of money, because I am not that interested in space. I think it's good. I think we ought to know about it. But we're talking about fantastic expenditures. We've wrecked our budget, and all these other domestic programs, and the only justification for it, in my opinion, is to do it in the time element I am asking.

Webb held his ground. He wanted the priority of NASA to be "pre-eminence in space," something that was much more open-ended than a lunar landing and would require unrelenting effort. At that suggestion, the president lost his temper: "By God, we keep telling everyone we're pre-eminent in space for five years and nobody believes it because they [the Russians] have the booster and the satellites. . . . You can't prove your pre-eminence unless you're on the Moon."[29] Returning to the subject nine days later, Webb insisted that the United States had to look beyond Apollo and "pursue an adequate well-balanced space program in all areas, including those not directly related to the manned lunar landing."[30] Kennedy remained unconvinced. What he wanted most was an administrator able to stick to a script.

The Mercury experiment came to a close with *Faith 7*, piloted by Gordon Cooper on May 15, 1963. Cooper stayed up for thirty-four hours, completing twenty-two orbits and earning the accolade of being the first man to sleep in space. (The mundane, apparently, becomes magnificent in a vacuum.) Cooper delighted the American public by saying a prayer while flying over the Indian Ocean, and then repeating it when he appeared before Congress after the flight. A delighted Brigadier General Robert Campbell of the California National Guard was moved to conclude that "there is no room for agnostics in America's space and missile program."[31]

The Soviets then trumped Cooper's space nap by repeating their magic rendezvous trick, this time with a twist. Valery Bykovsky went into space on June 14, 1963, followed two days later by Valentina Tereshkova. The fact that the two capsules eventually drew within less than 5 kilometers of each other was overshadowed completely by news of the world's first female cosmonaut. Reading from a script, she radioed back to Earth: "Warm greetings from space to the glorious Leninist Young Communist League that reared me. Everything that is good in me I owe to our Communist Party."[32]

Tereshkova spent three days in space, more than all the Mercury astronauts combined. In the process, she implied that the Soviet system was, in contrast to the American, enthusiastic about gender equality. Claire Booth Luce, writing in *Life*, thought that the mission was "symbolic of the emancipation of the communist woman. It symbolizes to Russian women that they actively share (not passively bask, like American women) in the glory of conquering space."[33] Jane Hart, wife of a Michigan senator, reacted angrily to the news, especially since she had passed the medical tests for astronaut qualification but had been ignored by NASA. She concluded that "the U.S. is a hundred years behind in using the full capabilities of women."[34] Another aspirant, Geraldyn Cobb, complained: "I find it a little ridiculous when I read in the newspaper that there is a place called Chimp College in New Mexico where they are training fifty chimpanzees for space flight, one a female named Glenda."[35] She wondered why she couldn't simply volunteer to take Glenda's place. Coming to Cobb's aid, Tereshkova remarked: "It is very sad the way American leaders have made a laughing stock of her. They shout at every turn about their democracy and at the same time they announce that they will not allow a woman into space. This is open inequality."[36] When asked about the possibility of women astronauts, von Braun quipped that "all I can say is that the male astronauts are all for it. . . . We're reserving 110 pounds of payload for recreational equipment." John Glenn smiled and said that he would embrace the idea "with open arms."[37]

The fuss over Tereshkova was unwarranted since this was another case of Russian bluff. Tereshkova was simply a tool of propaganda, just like poor Laika. Her flight indicated nothing other than that the Soviets were good at embarrassing the Americans. In fact, according to Alexei Leonov, "Most Russians believed that women should not meddle in what was regarded as men's work." The rest of the cosmonaut corps were angry at the way she demeaned their job, especially when she described flying in space as simple. In fact, Tereshkova spent almost the entire flight in a state of blind panic, though this was not revealed until 1990. "Korolev simply concluded that his spacecraft were unsuitable for women," Leonov added.[38] All the hoopla about the first female in space distracted attention from a rather important fact. Tereshkova was not a pilot. She was an ordinary garment worker, another characteristic that

Soviet propagandists milked for all it was worth. The fact that she had no piloting experience should have demonstrated (to those who cared to notice) that Soviet space capsules were controlled from the ground.

In contrast, Mercury proved that astronauts could fly in space, not just as passengers but also as pilots. A number of sticky situations had been avoided by the skill of the astronauts and their ability to cope in a crisis. Likewise, the opposite was true: some near crises were caused by astronauts making wrong decisions or by allowing their emotions to get the better of them. As long as the latter problem could be controlled, the revelation proved a boon to NASA, since astronauts were the agency's bread and butter. NASA needed to present them as men in control of their own destiny, not to mention that of the nation. The lunar quest depended upon a steady supply of heroes.

Out in the southeastern United States there's a weed called kudzu. Originally imported from Japan, it now covers over seven million acres. It kills other plants by smothering them under a solid blanket of leaves, by strangling stems and tree trunks, and by breaking branches or uprooting entire trees and shrubs through the sheer force of its weight. Once established, Kudzu plants grow incredibly rapidly.[39]

After Kennedy applied his fertilizer, NASA was a lot like kudzu. It spread over the entire United States, but was particularly virulent in the Southeast, where fat contracts kept conservative southerners quiet. NASA took over entire communities and sent its tendrils everywhere. Its healthy appearance obscured the fact that it was strangling the American economy. It defined a need—going to the Moon—which was supposed to have absolute importance. This importance meant that budgets could expand virtually unchecked and that NASA itself would not be subjected to performance appraisal. NASA was a parasitic weed that fed on the host economy, eventually seriously weakening it, but hiding the damage it did by encouraging a false confidence built on technological wizardry.

The biggest specimen of this virulent weed was the Manned Spaceflight Center in Houston, commonly known as Mission Control. Robert Gilruth had been administering the Mercury program from the Langley site in Virginia. He initially argued that Langley could easily be expanded, but Webb chastised him for even considering the idea. "Bob," Webb remarked, "what the hell has Senator Harry Byrd [of Virginia] ever done for you or NASA?"[40] Gilruth elaborated:

Mr. Webb . . . said, "We've got to get the power. We've got to get the money, or we can't do this program. And we've got to do it. And the first thing," he said, "we've got to move to Texas. Texas is a good place for you to operate. It's in the center of the country. You're on salt water. It happens also to be the home of the man who is the controller of the money." That was Albert Thomas. . . . The people weren't as gung-ho in Virginia as they were in Texas. They wanted us in Texas. They were thrilled to have the space program come to Texas.[41]

Senator George Smathers of Florida, who thought that everything should be centralized in his state, recalled a "big argument, big fight" with Johnson over the decision to put Mission Control in Texas. "Johnson tried to act like he didn't know. . . . It never has made sense to have a big operation at Cape Canaveral and another big operation in Texas. But that's what we got, and we got that because Kennedy allowed Johnson to become theoretical head of the space program."[42]

The deal would have made any Texan land developer proud. Humble Oil donated the site to Rice University, but kept the rights to the huge oil and gas reserves that lay below the surface. That allowed the oilmen to keep what they wanted, while getting a huge tax deduction for their charitable donation. Rice kept a portion of the land for itself, donated 1,000 acres to NASA, and sold the agency another 650 acres at $1,000 per acre. That meant a quick profit of $650,000 and a permanent tie to NASA. Everyone was happy, especially Johnson and Albert Thomas, who watched with delight as Houston expanded to absorb the NASA site, sending real estate prices skyrocketing.

One prominent part of the new facility was the media center, located in a two-story building just across the street from Mission Control. Twelve hundred phone lines were installed, and a four-hundred-seat auditorium was built in which press conferences could be held. During missions, briefings took place in the auditorium at least three times a day, more often if the mission did not go exactly to plan. Mission Control also provided live audio and video feeds where appropriate, with typed transcripts of conversations between astronauts and flight controllers supplied within hours.

NASA, of course, worked hard to maintain a positive image. The mega-corporation had an instinctive survival sense; it knew that public interest was its life support. The agency, for instance, gladly opened its gates to any Hollywood producer keen to inject a wholesome space theme into his film or television program. "Do you have any suggestions about how we clear a dog for access to KSC?" Walter Whittaker asked, in an internal memo, after the *Lassie* production team requested access.[43] Mister Rogers and Captain Kangaroo were also welcomed with open arms. In 1970, real Moon rocks were made available to Mayberry R.F.D., even though that seemed a rather obsessive quest for authenticity.

According to Amitai Etzioni, the agency invested millions into making sure that it always had a stable of friendly reporters willing to put inspiring stories in the press:

> NASA . . . organized seminars at schools of journalism in which they helped journalists get acquainted with the technical details. . . . And they would wine and dine them and take them to Houston, take them to launches . . . and they created quite successfully a whole cadre of reporters who specialized in space and who certainly the last thing they wanted was to terminate Project Apollo and the space program.[44]

In 1963, NASA offered the Columbia School of Journalism $393,000 to study techniques for improving the dissemination of space news and methods of training space journalists.* NASA's press kits were things of beauty, designed to get even the most hard-bitten reporter excited. The agency knew precisely how to make journalists feel welcome and important. Facilities were thrown open. Reporters were given access to a lot of beguiling gadgetry and occasionally got to pretend to be astronauts or mission control technicians by playing with real equipment. *Newsweek*, which for understandable reasons tended to be more objective about the space program than the magazines in the Time/Life stable, rued the fact that "reporters are under the temptation to function as rooters for 'The Team'—a role abhorrent to most newsmen."[45] One of the most enthusiastic converts was Walter Cronkite, who covered every space shot for CBS. Lynn Sherr, who worked for ABC, felt that Cronkite was "much more of a cheerleader [than a reporter]. A lot of the other reporters were reporting it a little more, I won't say objectively, but perhaps fairly. [They were] able to go down the middle rather than essentially being on NASA's side so much of the time."[46]

Politicians were coddled in much the same way as newsmen. "What the general run-of-the-mill congressman thought or how he an-

* The resultant controversy eventually caused the university to pull out of the deal.

172 THE SLEEP OF REASON

swered his people, depended a lot on his own situation," Webb remarked. Manipulating them was therefore a fine art. "Many did not
want to make a public commitment, but I've never known one where
you went into his district with an astronaut . . . and, at the end of the day,
after they had seen the tremendous response that this program drew
from the younger people in the district, they would say, 'This is fine, I'm
for it.'"[47] Brian Duff, the director of NASA's Public Affairs Division, recalled a particularly memorable instance of this technique, used on
George Mahon, chairman of the House Appropriation Committee:

> [Webb] found a time when Mahon was going back to his district and
> he gave him a ride. . . . He offered to give the congressman a ride in the
> Gulfstream and he then agreed to go with him and make some
> speeches in the district, and he took Gordon Cooper, the astronaut,
> with him. This was fairly early in the program. . . . In the back of the
> plane there were four seats, four big easy chairs that would swivel to
> face each other. Webb . . . put the astronaut in one of the other seats.

When they got to Lubbock, Mahon's district, Webb applied phase two
of the strategy:

> Webb and the astronaut . . . must have made a speech at every high
> school in Lubbock, and I think there were four of them. At the first one,
> Mahon introduced them and then he moved over and sat on the edge
> of the stage. By the time we got to the second one . . . he'd seen the ap-
> peal that Webb and especially Cooper had, and his chair began mov-
> ing across the stage, getting closer and closer to these two stars. By the
> last speech . . . the three chairs were together on the stage.[48]

Webb also knew how to appeal to a politician's base instincts—his
greed and lust for power. NASA was a huge pork barrel at which everyone wanted to feed. Representative George Miller, a Democrat from
California, was at first quite critical of NASA and the lunar mission.
Then, after the death of Overton Brooks, he took over the chairmanship
of the influential House Committee on Science and Astronautics and he
was suddenly converted to the NASA gospel:

> Our need to lead in space exploration is not merely a matter of sur-
> vival; it is not simply the result of a selfish desire for the yet un-

dreamed-of conveniences and luxuries which the mastering of space technology can create. It arises from a broader and nobler purpose which has existed in the hearts and minds of men since the first human thought occurred—the need to know, the need to grow, the desire for fulfillment of the ultimate destiny of mankind.[49]

As the journalist George Dixon remarked: "The attitude of many members of Congress toward the lunar landing program seems to boil down to this: 'I have no objection to a man going to the Moon, so long as he starts from my district.'"[50]

Forget science, forget about the need to explore—for most politicians the space program was simply a matter of money. Referring ironically to the way congressmen behaved like pigs at a trough, Dixon remarked: "When Princeton astronomer Martin Schwarzschild told the Senate Space Committee that the lunar project was the 'spearhead for a renaissance of America's pioneering spirit,' Senator Spessard Holland of Florida asked only how much of this moon money the National Aeronautics and Space Administration plans to spend in Florida."[51]

Less than a month after Cooper's flight, the Senate Committee on Aeronautical and Space Sciences held hearings on the future of the space program. Within the upper chamber, unease had arisen over the cost of the space program, while the justification for a lunar landing had been weakened by Kennedy's enthusiasm for cooperation with the Russians. The committee was particularly keen to hear the views of scientists outside NASA, in other words those who did not have a vested interest in the program. They called on Philip Abelson, editor of *Science* magazine, who revealed that an informal straw poll he had conducted among non-NASA scientists had yielded 110 against the manned lunar program, with only three in favor.* He elaborated:

Making man a part of the scientific exploration of space has two important drawbacks. It increases costs and it will probably slow down, at least for some years, the pace of getting valuable results. . . . The ar-

* On May 10, 1963, the Washington *Evening Star* remarked: "Of the 30 or 40 distinguished men who hold the prize professorships at Harvard, Yale, Columbia, Princeton, Chicago and California, not one is identified with—much less actively engaged in—the space program. . . . There is something seedy about using the space program as a cover for serious scientific enquiry."

gument has been made that putting a man in space will open vast frontiers of knowledge. No one has delineated any impressive body of questions which are to be studied. Rather we are reassured by the statement that "Man can meet the unexpected." . . . One of the things that has bothered me is that I read in the newspapers of estimates of the manpower that NASA is going to require in a few years and talk of one-third of the physical scientists being engaged in space exploration. This just makes no sense to me in terms of the opportunities in other fields of effort. I don't believe NASA should be given an unlimited hunting license. There ought to be some kind of limitations.[52]

"I believe," Abelson concluded, "that diversion of talent to the space program is having and will have direct and indirect damaging effects on almost every area of science, technology and medicine. I believe that the program may delay conquest of cancer and mental illness."[53] James Van Allen, whose research had benefited so greatly from the first Vanguard flight, thought that putting men in space was a waste of valuable rockets. He cited "overwhelming evidence that space science is best served by unmanned, automated, commandable spacecraft" and thought that the belief that space exploration had to be presented as a human adventure was "an insult to an informed and intelligent citizenry."[54]

The mathematician Warren Weaver agreed with Abelson that the space program was diverting valuable scientific research from more productive pursuits. He suggested another way to spend $30 billion:

It would provide a 10 per cent raise in salary, over a 10 year period, to every teacher in the United States from kindergarten to college, public and private; give $10 million each to 100 of the best smaller colleges; finance seven-year fellowships $4,000 per person per year for 50,000 new scientists and engineers; contribute $100 million each toward the creation of ten new medical schools; build and largely endow complete universities, with medical, engineering, and agricultural faculties for most of the 53 nations that have been added to the United Nations since its original founding; and create three more permanent Rockefeller Foundations.[55]

Commenting on Weaver's criticisms, one paper concluded that going to the Moon "could result in neglect not only of the real reason for the venture into space—man's desire for knowledge and control of his outer environment—but of our very economic lifestream."[56] What the scientists failed to realize, however, was that science was not important per se. Science and technology were useful to the government only to the extent that they could advance a political agenda and camouflage the shallowness of the space pursuit. Science was the servant of politics, not the other way around.

Many scientists, however, backed the program, if only for the money it brought. During the FY1965 budget debate, a group of eight eminent scientists made a very public display of support for NASA. The group included three Nobel Prize winners: Willard Libby, Joshua Lederberg, and Harold Urey. The latter argued that since instruments by themselves could only give "a crude imitation of human judgement and flexibility . . . situations are bound to arise in which the human performance is indispensable." Covered in all the major newspapers, the statement had significant impact. In an astonishing example of circular reasoning, the *San Francisco Chronicle* concluded that "the United States must attempt to reach the Moon because it is reachable by man and because it is inconceivable . . . that a great and powerful and scientifically advanced nation should turn away from the challenge. That the cost will be great is obvious, but inestimably far greater is the cost of not knowing what can be known."[57]

The mood in Congress had nevertheless changed radically since Kennedy set down his challenge in 1961. Legislators had been told that space was a frontier in the Cold War, a battle that had to be won. But the Cuban Missile Crisis in October 1962, when the Cold War adversaries came closest to nuclear confrontation, had, ironically, led to a thawing of relations with the Soviet Union. In the aftermath of the crisis, a demonstrable desire for peaceful coexistence had emerged. An atmospheric test ban treaty had been signed, as had the Hotline Agreement, a measure designed to ensure a better line of communication during times of crisis between the American and Soviet leaders. There was talk of a ban on weapons in space. Kennedy himself was repeatedly pushing the idea of cooperating with the Soviets in space. What, then, was the point of going to the Moon?

Fiscal conservatives began to sharpen their knives. Within the Senate, a move to impose significant cuts upon NASA was gaining momentum. David Scott, who submitted his application to join the astronaut corps in March 1963 ("principally because it was what you were expected to do"), remembers the great uncertainty which hung over the space program at that time. "NASA's budget was still terrible," he recalled. "Shortly after I put my application in, *Time* magazine ran an article with a picture of [Frank] Borman and [Ed] White sitting around in an office in Houston, alongside an article detailing the lack of funding for the Gemini program. The implication was that they were wondering whether they had made the right choice in giving up their Air Force careers."[58]

When Moon fever threatened to wane, the NASA public affairs office simply stepped up the tempo. In all of 1962, for instance, James Webb gave forty-nine speeches at various public gatherings, and Hugh Dryden another sixteen. In the *first half* of 1963 (when the Senate hearings were taking place), Webb delivered forty-two speeches and Dryden twenty-one.[59] Webb's speeches managed to blend the grand vision with stark reminders of the dangers of losing the space race:

> The nations of this world, seeking a basis for their own futures, continually pass judgment on our ability as a nation to make decisions, to concentrate effort, to manage vast and complex technological programs in our own interest. It is not too much to say that in many ways the viability of representative government and of the free enterprise system in a period of revolutionary changes based on science and technology is being tested in space.[60]

In April 1963, "Shorty" Powers of the NASA Press Office said that critics of the space program reminded him of "the people who delayed Christopher Columbus' voyage in the 15th century."[61] The press loyally echoed this theme. "A time of challenge always produces skeptics and nay-sayers," *Life* remarked. "Isabella of Spain had advisers who tried to talk her out of financing Columbus' voyage. But the bold spirits of that time did venture into the unknown. And they turned their age, already exciting enough, into an era of unprecedented exploration and discovery which changed the history of the world."[62] "Enterprises such as this are not to be judged by the ordinary criteria of scientists, economists or sociologists," the *Washington Post* decided. "Heroic enterprises move by their own laws, abide by their own rules and set their own precedents and when they are over, leave humanity with its knowledge multiplied, its future expanded, its horizons widened, its outlook sharpened and its hopes uplifted by a new sense of man's unending and unlimited possibilities."[63] Journalists also swallowed Webb's argument that he who controlled space also controlled the Earth. "If the Free World is to survive," the *Birmingham News* shouted, "we *must* not allow anyone else to gain the upper hand in space."[64]

NASA could depend on faithful friends within Congress. The great arm-twisting machine always delivered the necessary level of support. At one point during a crucial debate on NASA appropriations in 1963, Congressman H. R. Gross of Iowa noticed that some astronauts and their wives were present in the chamber. Smelling a rat, Gross asked Thomas if the arrival of the guests was "purely coincidental." Miller came to the rescue, offering the innocent explanation that the astronauts were in town to receive a medal and that they had expressed a desire to witness Congress in action. Gross was not convinced. A short time later, acting like loyal robots, Congress voted 335–47 to give $5.1 billion to NASA. Gross had earlier complained about the willingness of his colleagues to vote huge sums of money for NASA without seriously debating the wisdom of doing so. "I hope that if we do get to the Moon we find a gold mine up there because we will certainly need it," he concluded.[65]

The critics were defeated, but they did not go away. William F. Buckley wondered whether an appropriate epitaph for the space age might turn out to be "The Moon *and* Bust." "We do not need galactic bombast to prove to the people of the world that ours is a superior society to the Soviet Union's: we could, had we the imagination to do so, prove that point again and again, profoundly and convincingly, if we could only free ourselves from the inferiority complex to which we have been enslaved for years." He thought that Kennedy should simply tell the Soviets: "Very well, you have reached the Moon, but meanwhile here in America we have been trying, however clumsily, to spread freedom and justice."[66] *Newsweek* writer Edwin Diamond compared the whole venture to a Kwakiutl Indian potlatch ceremony in which rival chiefs try to outdo each other by tossing ever more valuable possessions into the campfire.[67]

A second line of argument was subtler. The worship of science had, over the previous half century, encouraged the creation of megalithic bodies that were supposed to deliver benefit to mankind but in fact served only themselves. Eisenhower had warned about this in his farewell address. Etzioni made the same point in his 1964 book *The*

Moon-Doggle. In *The New Priesthood*, Ralph Lapp drew attention to the power of the "technocrats," scientists who used their arcane knowledge as a source of power and also as a method of exclusion. The philosopher Lewis Mumford raised similar fears. "If the plight of the human race at the current moment is a desperate one," he wrote, "it is because during the last half century, Western man entered into a vast legacy of power without taking any prudent precautions against his own misuse of it." Ordinary people had allowed themselves to be brainwashed by a new generation of scientists who promised enormous benefit to all mankind, but in truth wanted only to pursue their own fantasies, some of which were dangerous to the world as a whole.

> Right up to this moment we have behaved like sleepwalkers without any control over our own behavior or any knowledge of where we actually are. Or any sense of what unplumbed chasms we are at this moment teetering above unable to shake off our old, beguiling dreams or conquering nature and unable to face the incredible nightmare of reality.[68]

The sleep of reason, Goya said, produces monsters. Mumford complained about "mega-machines" that acted like huge, voracious beasts controlling society. One example he cited was the Manhattan Project; another was NASA. These organizations assumed a life of their own. Their size meant that they wielded enormous control. Rather like human beings, they had an instinct for survival divorced from the service they provided. They became secretive, manipulative, and coercive in wielding power simply to ensure their own continuance.

Suspicion of technology was most prominent within the 1960s counter-culture. It would gradually coalesce and reach its most potent expression in the environmental movement. But the nature of its constituency limited its impact. Mainstream America did not want to hear hippies moan about the tyranny of technology. Nevertheless, behind the scenes, unease was building. Etzioni reckoned that about one-third of Americans were concerned by the rush of technology, and wanted a slower life.[69] He also felt that there was enormous social and political pressure preventing those who questioned the space program from speaking out. "I was . . . at Columbia, I had tenure very young and I considered that part of my job," he remarked. "Other people who would do what I did, could get into trouble; they would be considered traitors or they certainly wouldn't be employed by NASA . . . or get grants from it which was very important to researchers. But I was given this privilege and I considered it was my duty, my calling to . . . call the things the way I saw them."[70]

At times, Kennedy hinted at a vague fear of the ogre he had nurtured. On August 22, 1963, a reporter asked him whether it might be wise to militarize the space program, in order to ensure greater speed and efficiency. NASA, it was suggested, might be modeled on the Manhattan Project. Kennedy summarily rejected the idea, pointing out that "the fact of the matter is that 40 per cent of the R&D funds in this country are being spent for space, and that is a tremendous amount of money and a tremendous concentration of our scientific effort." In an effort to defend himself, the president merely revealed the extent of the monster he had created. "I think the American people ought to understand the billions of dollars we are talking about, which I believe a month ago was mentioned as a great boondoggle." It wasn't a boondoggle, he insisted. "I think it is important, vital, and there is a great interrelationship between space, military, and the peaceful use of space. But we are concentrating on the peaceful use of space which will also help us protect our security if that becomes essential."[71]

But what did this mean? The ordinarily articulate Kennedy seems to have had his sense of reason strangled by the tentacles of technocracy. The statement encapsulates just how muddled his space initiative had become. In order to please conservatives he had to continue to present the program as an effort to strengthen the nation's defenses, even though instinctively he did not want to extend the Cold War into outer space. In order to please the dreamers, he had to talk up the new frontier which would "benefit the people of all nations." Yet, two years after his famous speech, America was neither safer, nor happier, nor more inspired because of what had been achieved in space. The reference to "billions of dollars" suggests a frustration with the fact that all Kennedy had managed to do so far was spend a lot of money.

In late October, Khrushchev made what Kennedy called some "rather Delphic" comments regarding the Soviet space program, which some commentators interpreted as an indication that the Russians wanted to opt out of the space race. Hints to this effect had been made at various times throughout the summer. On this occasion, Khrushchev told reporters: "It would be very interesting to take a trip to the Moon, but I cannot at present say when this will be done. We are not at present planning flights by cosmonauts to the Moon."[72] The newspapers, accustomed to telling readers that the Soviets were far ahead in the space race, could not contain their astonishment. Khrushchev, alarmed at the reaction he had provoked, immediately backtracked. "Give up such hopes once and for all and just throw them away," he told Americans. "When we have the technical possibilities of doing this and when we have complete confidence that whoever is sent to the Moon can safely be sent back, then it is quite feasible. We never said we are giving up on the lunar project. You are the ones who said that."[73]

The issue underlined Kennedy's dilemma. He would have loved to be relieved of his commitment to go to the Moon. But, in order to sell the program, he had been forced to present it as much more than a race. The justifications he had provided did not theoretically depend on the Soviets remaining interested, but Soviet interest nevertheless validated them. As he admitted on July 17, 1963, "the capacity to dominate space, which would be demonstrated by a Moon flight . . . is essential to the United States as a leading free world power. That is why I am interested in it and that is why I think we should continue." When the possibility of the Russians pulling out resurfaced in October, Kennedy insisted that the United States would not alter its commitment.

> The space program we have is essential to the security of the United States, because, as I have said many times before, it is not a question of going to the Moon. It is a question of having the competence to master this environment. . . . We have no idea whether the Soviet Union is going to make a race for the moon or whether it is going to attempt an even greater program. I think we ought to stay with our program. I think that is the best answer to Mr. Khrushchev.[74]

By October 1963, the Kennedy space program was a dog's dinner. He had made the great pledge to land on the Moon by the end of the decade. He had repeatedly referred to Soviet and American space efforts as a race, a race that the United States had to win. But while the presence of the Soviets in that race made life difficult for him, given their tendency to trump NASA's every move, Kennedy could not do without that presence. Without the Soviets in the race, there was little point in spending all that money. The Soviet presence validated everything that Kennedy had said about the intrinsic importance of space. That is why one notes a tone of desperation in the Kennedy responses quoted above. He was painfully aware that without the Russians, the space quest would lose its point. He was desperate to find a way out of the bizarre pledge he had made, but the Russians would not cooperate.

On November 22, 1963, Kennedy was due to address the Texas Democratic State Committee in Austin. In the middle of his wide-ranging speech, he was supposed to reiterate the importance of space:

> The United States of America has no intention of finishing second in outer space. . . . This is still a daring and dangerous frontier; and there are those who would prefer to turn back or to take a more timid stance. But Texans have stood their ground on embattled frontiers before, and I know you will help us see this battle through.[75]

The speech was never delivered. While his motorcade was passing through Dallas, Kennedy was fatally wounded by an assassin's bullet. His assassination gave the space program an importance he never intended.

Kennedy, it appears, dreaded the monster he had created. The exploits of the Mercury astronauts brought him enormous popularity, but also a massive financial commitment. He had encouraged the growth of NASA into a mega-institution that devoured money faster than any other federal program. Recognizing what effect this monster would have upon his domestic plans, he dearly wanted to curtail the space race, but did not know how. It is reckless to speculate upon what might have happened if Kennedy had lived. But what is clear is that the coalition the president had created in 1961 had begun to unravel. It is also clear that Kennedy would have welcomed a way out of the space race, if not through cooperation with the Soviets, than perhaps through restrictions imposed by a penny-pinching Congress. But none of that happened. After the assassination, the space program became a homage to Kennedy and, as such, untouchable. Shortly after *Apollo 11* landed on the Moon, an unknown person left a simple message on Kennedy's grave at Arlington National Cemetery. It read: "Mr President, the Eagle has landed." One of Kennedy's most stunning achievements was the way he managed to convince Americans that he actually cared about space. Eulogies to the "space president" occupy acres of print. Some of the most perceptive scholars have been fooled by his words at Rice University. While the Kennedy image has tarnished in so many other respects, his devotion to the space program has gone largely unquestioned. "More than anyone else Kennedy had created an aura of romance around space," David Scott remarked, more perceptively than he understood. "He had been the greatest single supporter of NASA. More than anyone else, he had captured the public imagination. His enthusiasm and absolute determination convinced many Americans that he would win the race to the Moon."[76] One thing is certain. Kennedy was much more valuable to NASA dead than he ever was alive.[*]

On November 22, 1963, Kennedy was supposed to have featured in the first live broadcast transmitted across the Pacific, via the new communications satellite Relay 1. But his opening remarks were canceled due to his assassination. Four days later the satellite beamed film of Kennedy's funeral to millions of people across Asia and Europe, including the Soviet Union. Never before had world events been so immediate. The telecast of Kennedy's funeral around the world should have demonstrated that the United States was far ahead in the kind of space technology that really mattered. It did nothing of the sort. The United States would go to the Moon for the sake of the president who never really cared.

---

[*] It only remains for the conspiracy hounds to sniff a plot, and they have, as the Internet attests. Here's the scenario: NASA, frightened that Kennedy would back out of his pledge, had Kennedy killed. The plot links Bobby Baker, his friends in the underworld, and Michael Paine, an acquaintance of Lee Harvey Oswald who worked for Bell Aircraft. Bell was run by Walter Dornberger, a friend of von Braun. The tests on the "magic bullet" were apparently carried out by Thomas Canning, a NASA engineer.

10

Lost in Space

The image of America in the 1960s is often one of turmoil (assassinations, war, protest) mixed with hedonistic fun (sex, drugs, rock and roll). But what analysts often fail to appreciate is that, for so many Americans, the space adventure was an all-consuming distraction that diverted their attention away from the domestic and international crises usually associated with the decade. Instead of looking at the problems around them, many Americans cast their gaze skyward. America was lost in space.

Television schedules were packed with space themes. Hollywood offered a romantic comedy (*I Dream of Jeannie*), two slapstick sit-coms (*My Favorite Martian* and *It's About Time*), a corny family adventure (*Lost in Space*), a "serious" bit of science fiction (*Star Trek*), and a variety of space cartoons, including *The Jetsons*. The exploits of NASA provided welcome verisimilitude to the space stories peddled by the Hollywood film industry. The boom in space films that started in the early 1950s continued through the '60s and '70s. Most were B-movies deploying the well-worn theme of astronauts encountering something nasty in outer space, as with the truly execrable *Phantom Planet* (1962), *Assignment—Outer Space* (1962), and *Wild, Wild Planet* (1967). Chimps in space provided enormous opportunity for comedy, as in Disney's *Moon Pilot*, starring Tom Tryon and a since-forgotten monkey. Chimps of a decidedly more sinister sort figured in *Planet of the Apes* (1968) and its endlessly diminishing sequels.

Missing from the genre were heroic stories of successful space missions, perhaps because NASA itself had a monopoly on that plotline and because space travel is pretty boring unless something goes wrong. This nevertheless caused a problem for NASA, since movie producers regularly asked for technical assistance and location shots, but the agency did not want to be associated with films about space disasters. A case in point was Frank Capra's *Marooned*, universally panned on its release in 1970. While the film was still in its planning stages, the PR supremo Julian Scheer wrote, in an internal memo: "It would be better for the agency's standpoint if the picture were never made." A request to use the NASA logo in establishing shots was refused.[1]

Rock music and space travel grew up together. There's hardly a baby boomer around who doesn't remember listening to The Ventures' "Telstar" on a cheap transistor radio.* The space theme was expressed in virtually every pop genre from Bill Haley's "Rocking Chair on the Moon" (1952), to David Bowie's "Space Oddity" (1969), to Donovan's "Intergalactic Laxative" (1973), to The Byrds' "Mr. Spaceman" (1966). In addition to these songs, quite a few others were written, but fortunately never recorded. NASA was inundated with unsolicited mail from amateur songwriters moved to write theme songs for the lunar quest, including "Hail to the Astronauts," "Bless Thou the Astronauts," and "The Hymn of the Astronaut." Submissions of this sort all received the stock reply that NASA had no plans to adopt an official song.[2]

For a while, it seemed like every hotel, motel, bar, restaurant, shopping center, or strip mall had rocket, planet, space, or Moon in its name. At Christmas, Santa brought space suits and pens that could write upside down. For 15 cents and few cereal boxtops, children could get a space ring or a miniature Mercury capsule, or a map of the Moon. In 1968, space toys ranked top in popularity, with war toys a distant second. Coca-Cola started the Space Education Club—for one dollar and a handful of bottlecaps, any child could get "a Space Explorer Club cloth patch, an official Apollo patch and a personal letter of congratulations from an astronaut."[3] Shell Oil gave its customers commemorative plastic coins adorned with the faces of the astronauts—these could be displayed in a simulated walnut presentation case costing only $5.99. The "Walking Billy Blastoff" was an astronaut who came with seven pieces of paraphernalia, including a space car that looked suspiciously like a 1959 Chevy Bel Air. In 1970, GI Joe took off his military uniform and donned a spacesuit. He came complete with a space capsule and retailed for just $8.93 at Toys R Us. In November 1969, the retailer took out a three-page ad in the *Washington Star*. Two of the pages were devoted completely to space toys.

The NASA archives contain thousands of letters from young boys (and a few girls) who wanted to be astronauts. "Dear Mr. Kennedy," one

* The song was originally released by The Tornadoes, but most people remember the version by The Ventures.

confused young man wrote. "If you need anyone to go to the Moon in one of those Liberty Bells I will go if you will fix the crack so no Moon men can get in."[4] NASA had a standard reply for wannabe astronauts: "Thanks for your admirable devotion to your country, but for now the best thing you can do is work hard at school, especially in mathematics and science." Disappointed boys turned in droves to imaginary space travel, to the great delight of the model rocket industry. By 1970, its profits rose to $18 million a year. Ninety percent of the rockets were bought by boys between the ages of 9 and 14. The young rocketeers, rather like NASA, eventually decided that "manned" flight was much more interesting. Before long, boys started stuffing grasshoppers, lady-bugs, and mice into their nose cones.[5]

Since NASA went out of its way to promote the Boy Scout image of the astronauts, it's not surprising that the agency also cultivated active links with the scouting association. A merit badge in space exploration was created in 1966; the pamphlet outlining the requirements carried a testimonial from Webb. In order to earn the badge, a boy had to demonstrate a knowledge of the history of space exploration, be able to identify a range of rockets and space vehicles, build a model of a launch vehicle, and give a talk on the value of the space endeavor. Between 1965 and 1969, over eight thousand boys earned the badge.[6] The organization was also allowed to mention the names of astronauts with a scouting background in its promotional literature.

The race for the Moon inspired a love for all things artificial. On April 9, 1965, the National League Astros (the team formerly known as the Colt 45s) played their first game in the high-tech Astrodome. The Mercury astronauts were guests of honor. When it was found that grass would not grow in the covered stadium, it was replaced with plastic Astroturf, which never needed to be cut. Before long Americans were digging up their own lawns. Teflon was applied to pans and coats were fastened with Velcro. For most of the 1960s, it appeared that nylon, rayon, and polyester would replace cotton as the fabrics of the future. The space blanket could be slipped in one's back pocket yet supposedly provided as much warmth as a down comforter. Space Jackets were made of Mylar—they weighed less than eight ounces, never wrinkled, and could, the advertisement boasted, be stuffed inside a Coke can. For breakfast, Americans shunned fresh orange juice and drank instead Tang, the powdered citrus drink the astronauts took into space. Suddenly, it seemed, everything from coffee to soup to potatoes came freeze dried—just like the astronauts enjoyed. Space Food Sticks, a high-energy bar that came in peanut butter, chocolate, and butterscotch flavors, were marketed around the world after they debuted on the *Apollo 8* voyage to the Moon. "Food sticks could be a lifesaver in Space," the makers, Pillsbury, boasted in a full-page ad in the *Washington Post*.[7] They looked like a Tootsie-Roll, but tasted like a cotton rag. The only thing going for them was the connection to NASA, so when Apollo ended, so did they.* Space foods tasted terrible, but that didn't matter since they were so darned convenient. Besides, eating food that tasted like the lining of a lunar boot was a way of vicariously enjoying the thrill of space travel.

Tang, the synthetic orange juice, was widely assumed to have been pioneered by NASA, but in fact, it hit the supermarket shelves in 1957, before Sputnik. The space agency adopted it in 1965 on the Gemini flights, much to the delight of General Foods, the manufacturer. From 1965 to 1973, it was impossible to find an advertisement for Tang that did not mention the astronauts. This annoyed the Florida citrus growers, who thought that NASA was promoting the idea that Tang was preferable to real orange juice. "Florida citrus people," wrote the *Orlando Sentinel*, "would like to see the orange-flavored synthetics on just one more flight to the Moon—one way and with no astronauts aboard."[8]

The commodification of the space race should come as no surprise, given that the United States was *the* consumer nation. A healthy symbiosis existed between the producers of space-related products and NASA; those producers enjoyed the promotion the space quest provided (what would today be called cross-marketing), while NASA appreciated the way space age products imbedded their endeavor into everyday lives. There was no grand marketing plan, but rather an ad hoc effort by hundreds of producers to latch on to the space race as a way to demonstrate how modern they were. Space, in which new things seemed to be happening everyday, became synonymous with progress. But the association did not last. The space quest seemed at one point perpetually new and, as such, a hugely valuable, never-fading symbol of vitality. In fact, since going to the Moon was a passing fad, space as a marketing scheme had a limited shelf life. When the excite-

---

* Today, there's a Space Food Stick Preservation Society whose members are also enthusiastic supporters of a mission to Mars.

ment waned, space-related products were revealed as sorry imitations of the real thing. It is nearly impossible to find any of those products today, nor indeed any use of the cosmic quest as a marketing angle.

For many people, the space program was an island of harmony and hope in a sea of madness. The 1960s was a bewildering era, given the wars, assassinations, violent protests, free love, and drugs. In contrast, the astronauts were clean-cut American boys, old-fashioned heroes doing incredibly modern things in a down-home way. Brian Duff, whose job was to sell space to the American people, understood that without their support not a single NASA rocket would ever have flown:

> You can go back and say, "Did it really matter that we had public support? Did it make any difference? Wouldn't the old style have worked just as well?" We had Lyndon Johnson, what do we need public support for? We had George Mahon, we had two or three key people, why did we need public support? Webb would have said, "It's the public support that lets these other people give us what they want to give us. It's George Mahon knowing that back in Lubbock, his constituents are quasi-convinced that the space program is a good thing." . . . Our interest was to keep a drumfire of positive public attention on the program and never let up on the visibility of it. There was a sense of positive movement, that the whole thing was rolling.[9]

For millions of Americans, the space program was a wellspring of success—proof (at a time when proof was needed) that America as an ideal still worked.

The half million people directly involved in building the machinery that would take an American to the Moon had special reason to love the space program. For a craftsman at General Dynamics in San Diego, 1960 and 1961 were tough years. He was lucky if he worked three weeks out of four. Then came the big NASA budgets, and not only was employment dependable, but paychecks ballooned with overtime. Loyalty to the program was not, however, exclusively materialistic. Peggy Hewitt had worked at Grumman Engineering for many years, doing her time building ferocious weapons. "I guess my main motivation for wanting to work on the [Lunar Module] program was to get away from war. . . . *Really.* Even before Grumman got the contract I started maneuvering, manipulating. You can understand that feeling if you'd worked on machines of destruction for twenty years. It's about time to stop, isn't it?"[10]

Glynn Lunney, chief of the flight director's office throughout most of the Apollo lunar program, believes that those directly involved thought they were doing something special for America. "We could personally do our best to assure that our part of this national scene was going to go well," he recalled. "I think it gave us a sense, perhaps, of insulation from the emotional fallout from all these other things that were going on, the frustration, the lack of control, the stressing part of it. . . . We had this thing we could do, so it kept us together, and it was a little bit like, when we did our thing, we were on a little island and all around us were all these terrible thunderstorms and hurricanes and tornadoes and earthquakes."[11]

The actual popularity of the space program among the American public is difficult to measure. True, the manifestations of the space adventure in popular culture were ubiquitous. But just because Americans had the Moon on the mind did not necessarily mean that they wanted to go there. In 1961, when the Gallup organization asked whether it was important for the United States to be ahead of the Soviet Union in space exploration, 73 percent said that it was. When the same group was asked whether the United States should spend up to $40 billion to go to the Moon, 58 percent gave an emphatic "no."[12] In the summer of 1965, one-third of the nation wanted to cut the NASA budget, while only 16 percent wanted to increase it. By January 1969, the proportion favoring cuts had risen to 40 percent, while those who wanted increases fell to 14 percent. A close examination of American behavior reveals that while NASA exploits were popular, their cost was not. When pollsters removed economic references from questions about the popularity of the lunar mission, support shot up. On the other hand, at only one point, in October 1965, did more than 50 percent of Americans approve of the amount being spent to go to the Moon. Immediately after the lunar landing, an event that inspired massive rejoicing, only 53 percent agreed that the result justified the expense.[13] The fickle nature of American support demonstrates why NASA invested so much money and effort into public relations.

Twenty-two months separated the last Mercury flight from the first Gemini. Since the space program was primarily entertainment, that delay did not bode well for President Johnson, who understood more than anyone how a political career could be built on a launchpad. In the meantime, more serious problems at home in 1964 began to distract attention from the space adventure, in particular the increasingly strident discontent of African Americans. There was also the matter of a presidential election. While LBJ was a shoo-in, his challenger, Barry Goldwater, did make disturbing noises about national defense, and the part that NASA should play. On the domestic scene, Johnson's Great Society would compete with NASA for a share of the federal purse. Meanwhile, problems were brewing in Vietnam, another sinkhole for federal funds.

The Soviets kept interest in space alive, but not quite like Johnson would have wished. For them, Gemini represented a threat to their illusion of superiority. They planned to jump from a one-man capsule to a three-man one, but Soyuz would not be ready until 1967, by which time the Americans would already have moved on to their three-man Apollo. Desperate to maintain the illusion of a lead, Khrushchev pressured Sergei Korolev to come up with another trick. Ever the dutiful servant, the chief designer took an ordinary Vostok, stripped the interior off all but the essentials, stuffed three diminutive cosmonauts into it, and called it Voskhod. Korolev's deputy, Vasily Mishin, later remarked:

> Fitting a crew of three people, and in spacesuits, in the cabin of the Voskhod was impossible. So—down with the spacesuits! And the cosmonauts went up without them. It was also impossible to make three hatches for ejection. So—down with the ejection devices. Was it risky? Of course it was. It was as if there was, sort of, a three-seater craft and, at the same time, there wasn't. In fact, it was a circus act, for three people couldn't do any useful work in space. They were cramped just sitting! Not to mention that it was dangerous to fly.[14]

Instead of ejecting just before touchdown, as had previously been the practice, the cosmonauts had to ride their craft to the ground. Two retrorockets were added to slow their fall, but that still meant an abrupt return. The three sardines spent what has to be the most uncomfortable twenty-four hours in space, falling back to Earth with a bump on October 13, 1964. The trick nevertheless worked. The world's press assumed that the Soviets had yet again forged ahead.

Gemini was a workhorse program designed to prepare for the particular challenges that would present themselves on the way to the Moon. Four main capabilities had to be developed. The first was the ability for two vehicles to find each other in space, and then dock. Second was that of space walking or, in NASA jargon, extra-vehicular activity (EVA). Third, NASA had to develop the ability to carry out precision landings on Earth by means of guided reentry. The fourth requirement was the ability to conduct long-duration missions, an area in which the Russians were far ahead. NASA wanted to study the physical and psychological effects of an extended stay in weightlessness, with two astronauts stuffed inside a painfully small capsule.*

After suborbital tests of *Gemini* and its rocket, the first manned version blasted off on March 23, 1965, with Gus Grissom and John Young on board. The main lesson to be learned from this mission was the ability of the astronauts to maneuver *Gemini* into higher and lower orbits. This was of course essential in future missions, when spacecraft would have to dock with each other. Some serious cutting-edge science was also scheduled, to keep the astronauts busy, to provide NASA with scientific credibility, and to bring enormous benefit to mankind. On this occasion, one of the more important experiments was designed to measure the effect of zero gravity on sea urchin eggs. Unfortunately, an equipment malfunction wrecked the experiment—or so NASA claimed. Back on Earth, the blow to science was massive.

One unscheduled experiment took the dour technicians at Mission Control by surprise. Urged on by Wally Schirra, Young stuffed a corn beef sandwich in his spacesuit and offered it to Grissom when the two were circling the Earth. Faced with a choice between a meal of beef flavored toothpaste or a deli sandwich, Grissom understandably chose the latter, in the process causing the interior of the cabin to fill with floating shreds of pickled beef. The experiment demonstrated that certain foodstuffs do not perform well in a weightless environment. The public loved the stunt, proof positive that the brave astronauts were also fun-loving guys. But NASA, afraid that a multimillion-dollar mission might end in disaster if a bit of beef fouled some sensitive equipment, was, perhaps understandably, furious. So, too, was Congress. George Mueller, NASA deputy administrator at the time, had to promise a congressional hearing that steps would be taken "to prevent the recurrence of corned beef sandwiches on future flights."[15] Strict controls were placed on the personal items astronauts could henceforth take into

---

* For a hint of what that might have been like, choose someone with whom you think you might get along, get in one of those tiny cars (such as a Suzuki or a Kia), and stay there for two weeks, in the meantime performing all the bodily functions necessary to human survival.

space. For years afterwards, delicatessens in Cocoa offered a sandwich called a "Gus Grissom."

The corned beef controversy highlighted the contrasting cultures of those who built the rockets and those who flew in them. Both were indispensable, but they came from different worlds. The rocket jocks were intelligent men, but they had been selected for the job because of their individual courage. They enjoyed living on the edge, piloting super-fast planes or cars to their limits. Fulfillment came from spitting in the face of death. Practical jokes very often came with the package; they were a method of asserting individuality in a space program that prized uniformity and predictability. Stunts allowed these men to exert their will over their machines, thus demonstrating that the space program wasn't run by robots.

The scientists and engineers, in contrast, thought primarily of percentages. They regularly pushed the existing technology beyond the margins of prudence, and therefore were painfully conscious of the need to remove, as far as possible, unpredictables. While they were not automatons, they were mechanically minded, and in love with precision. The slide rule was their talisman. For them, machines were beautiful, predictable, and clean, while human beings were temperamental, complicated, and worryingly capricious. They did not appreciate it when stray shreds of corned beef, a risk that no computations could predict, polluted their perfect machines.

Space travel required a precise mix of romance and cold, hard science. The science made the rockets fly, while the romance ensured that taxpayers would foot the bill. The PR people liked the astronauts' fun-loving, "aw, shucks" attitude, but the engineers did not. At Mission Control, emotion was an inconvenience. Human beings were the greatest uncertainty; robots would have made the task so much simpler. When selecting astronauts, NASA had to find men who would appeal to the public but who, in training and on missions, would behave like automatons. As a result, astronauts were split personalities held together by a spacesuit.

Keeping everything under control were the senior managers, men like Robert Gilruth, Hugh Dryden, Robert Seamans, George Mueller, Max Faget, and James Webb. As much as the astronauts got the credit for being the heroes of the space program and, as much as Neil Armstrong, will always be remembered as the first man to walk on the Moon, the lunar mission was in truth a triumph of bureaucracy. The technology existed, at least on paper, before the venture began. What was needed was to take that technology, pull it into coherent form, and transform it into a system capable of putting a man on the Moon. Huge resources of labor, raw materials, and industrial infrastructure had to be coordinated precisely so that the quest could continue without hitch.

The need for a human side meant that astronauts spent a great deal of time out among the public, talking to civic groups and pressing the flesh. They would also regularly visit the factories where their machines were built. If the worker who welded the induction tubes of a Saturn rocket had in the back of his mind Gordon Cooper's smile, he might put that extra bit of care into his job—perhaps the difference between life and death. If going to the Moon was mainly a problem of management, managing the image was probably the most relentless task.

By the time the second and third crop of astronauts had been chosen and trained, the novelty of the astronaut had begun to wear off. "The Mercury guys' reputation for somewhat high living followed us around to some extent," David Scott recalled, "but the complexity of the Gemini missions meant that we had less time than they did for extra-curricular activities."[16] NASA also kept closer tabs on the astronauts' behavior. When John Young began a highly visible affair with a rather voluptuous woman, he was called into Gilruth's office for a fatherly chat. A short time later, he married the woman. "We were all selling our services and NASA was the only buyer," Walt Cunningham confessed. "Holding things together was part of the price each of us was willing to pay for that ticket into orbit. All of us had been willing to suppress personal feelings, desires, even a home life, good or bad, to keep the gods of space happy."[17] Not everyone, however, obeyed the rules. When Duane Graveline, who had been chosen as one of first scientist-astronauts, found himself in a messy divorce, NASA demanded his resignation. The first seven astronauts who got divorced never again got a flight.

One striking feature of the space race was the contrasting emotions of those who played a prominent role. The politicians had their own set of reasons for going to the Moon; the astronauts had an entirely different and more personal set. Politicians—Kennedy, Khrushchev, and Johnson—thought mainly of the race. The astronauts welcomed the opportunity to run in that race, but it was not their prime motivation. "For those in the Kremlin and in Washington, space was a fierce battleground for superiority not just in technology but in ideology," Alexei

Leonov felt. "But I . . . had no time or desire to contemplate such conflict. As far as I was concerned I was there to prove to my fellow man what human beings were capable of."[18] Every once in a while, astronauts and cosmonauts demonstrated that, even though their race was trivial, there was something magnificent about it.

Leonov very nearly became a casualty of that race when he went up in a capsule designed more with propaganda than safety in mind. In March 1965 he left his Voskhod capsule to embark upon history's first space walk. The maneuver nearly ended disastrously. The Soviets relied upon an air lock chamber through which the EVA astronaut had to pass before starting his space walk. What they did not foresee was that the difference in pressure between the air lock and the outside would cause the astronaut's suit to inflate. While the walk itself went well, Leonov soon began to resemble the Michelin Man, making movement of his joints increasingly difficult. Even more worrying was that he could no longer fit through the hatch of the air lock. He was forced to resort to the highly dangerous but inevitable measure of releasing some of the pressure in his suit while at the same time trying to maneuver himself back into the capsule. The sheer strain of the effort almost killed him.

The situation then went from bad to worse. When it came time to return to Earth, the retro-rockets wouldn't fire. The cosmonauts were faced with the possibility of endlessly orbiting the Earth and eventually dying in space. As a last resort, they fired their landing rockets, a highly imprecise technique, but still sufficient to jolt them out of orbit. They came down pinned between trees in a snowy forest, and spent a very uncomfortable forty-eight hours awaiting rescue, while hungry wolves circled the capsule. The entire mission was a fiasco, but the world's press, accustomed to reporting Soviet brilliance, saw no need to change their ways. Leonov's EVA was profusely praised, as was the "first manually controlled re-entry," with no mention that the latter was an act of desperation. Some perceptive journalists were beginning to question Soviet proficiency, but they didn't get much attention since everyone seemed addicted to the idea that the Russians were supreme. Korolev, shaken by the malfunctions of Voskhod, decided not to try again with the makeshift craft. But, since Soyuz was two years away from being ready, that meant that the Americans would have space to themselves. The Gemini plan called for a mission every two months, regular as clockwork. Russian magic had run out, at the very time that American efficiency was reaching its stride.

The Americans were certain that they could improve upon Leonov's EVA. *Gemini 4* followed on June 3, with James McDivitt and Edward White on board. White simply opened the hatch of the capsule and maneuvered himself out. This was simple, but it meant that McDivitt was also in effect walking in space, since the hatch was open for the entire duration. The short walk went well, with White maneuvering by using a high-tech gun that fired bursts of pressurized oxygen, but when he tried to return to the capsule he found that it had not been designed for ease of reentry. A tense few moments passed as White's fate hung in the balance. The EVA was technically a success, but NASA was painfully aware of how close the space program had come to its first fatality. Instead of tragedy, Americans got triumph, subsequently celebrated in "The Walk of Ed White" by the sickeningly wholesome pop group Up With People.

Gemini 5, with Gordon Cooper and Pete Conrad on board, followed on August 21. The purpose this time was to test the guidance and navigation systems exhaustively over a flight of 120 orbits. That flight went well, despite problems with newly developed fuel cells, used in place of conventional batteries. The next great challenge was for two craft to maneuver into close proximity—intentionally, not serendipitously, as the Russians had done. The original plan called for *Gemini 6*, with Wally Schirra and Tom Stafford aboard, to dock with an Agena satellite launched around the same time. Unfortunately, the Atlas rocket that was supposed to take the Agena aloft on October 25 exploded shortly after takeoff, forcing a cancellation. NASA then combined the mission profiles of *Gemini 6* and *7*, the latter being assigned a long-duration flight of two weeks. *Gemini 7* was launched first on December 4, with the Schirra-Stafford mission delayed until the 15th due to problems with the Titan II rocket. In a short flight of a little more than 24 hours, Schirra and Stafford quickly closed on Frank Borman and James Lovell in *Gemini 7*, coming as close as one foot away.

For anyone who cared to look closely, it was now clear that the U.S. lead in the space race was huge. During 1965, the Soviets had managed only Leonov's hapless Voskhod flight. The Americans, in contrast, had managed five Gemini flights in a space of just eight months, in the process performing a space walk and a genuine rendezvous. They'd also broken the endurance record twice, first with a flight of eight days, and then with the Borman/Lovell flight of two weeks. Smaller, but still hugely important achievements such as the refinement of fuel cells demonstrated that the Soviets were hopelessly behind.

But then came a near disaster. Docking with the Agena became the responsibility of *Gemini 8,* with its crew of Neil Armstrong and David Scott. This time, the Agena went into orbit without mishap. The *Gemini* launch followed a short time later and the docking went flawlessly, whereupon all hell broke loose.

As the combined craft passed over China, it started violently spinning and rolling. Having prepared for such a problem in the flight simulator, Scott and Armstrong went through the prescribed remedial responses, only to find that the spinning worsened. Assuming that the culprit was the Agena, they jettisoned the larger craft, but the spinning quickened. "We have a serious problem here," Armstrong told Houston.[19] By this stage the capsule was spinning at the rate of one revolution a second and the astronauts were close to blacking out. "We had lost control of *Gemini 8* almost completely," Scott recalled. "Our roll was speeding up. Neil was still working his hand controller to try to slow us down, but he was getting no response. It was like being on a theme park ride . . . except theme park rides don't spin so fast or for so long—if they did, too many of the passengers would black out."[20] "I was getting gray around the edges," Armstrong later confessed with typical understatement. "It was getting uncomfortable."[21] Looking back, Scott can't actually remember being afraid they were going to die. "The truth is that we were trained to perform at our highest peak in such situations. Emotion did not come into it. Everything was happening so fast and we had to find a solution. That took every ounce of our energy, every effort of our concentration."[22]

The media quickly got wind of the crisis and went into emergency mode, breaking into scheduled television broadcasts. At ABC, this meant delaying the broadcast of *Lost in Space.* Disgruntled viewers overloaded network switchboards with complaints.

NASA's standard procedure was to install "squawk boxes" in the homes of the astronauts on a mission, so that their families could listen to a live feed of the chatter between the capsule and Mission Control. Shortly after the problem was reported, someone decided to pull the plug on the transmissions. A distraught Jan Armstrong called Mission Control for news of her husband and got some second-rate official spouting platitudes. She asked for permission to come to the flight control center, but was refused. "The excuse they gave us about not letting us in . . . was that they couldn't face us if something went wrong."[23] She and Lurton Scott had simply to wait like dutiful wives.

Seamans, the NASA deputy administrator, just happened to be attending a Wright Brothers dinner in Washington, along with fifteen hundred guests. The dinner, an annual affair, brought together the leading lights of the aviation and aerospace industries. Seamans was supposed to sit at the top table with Vice President Hubert Humphrey. "When I drew up to the hotel . . . a gang of NASA people . . . descended on me and swept me into a room and briefed me on what was going on. . . . I was told that we were in deep trouble. I felt I had to notify everyone at the dinner that we had this problem, 'cause I felt it would be leaking out because the networks were already carrying it."[24]

Meanwhile, out in space, the rate of spin increased. "My peripheral vision was beginning to fade," Scott recalled. "I was getting tunnel vision. G-forces were throwing our heads away from the axis of rotation." They were running out of options, and would soon be incapable of taking action. "Everything was happening so quickly. We had only seconds left before we lost consciousness. We were out of range of any tracking station. We had no time to consult Mission Control, anyway. Neil had a very tough call to make."

"All we have left is the re-entry system," Armstrong gasped. "I knew he was right," Scott recalled. "Our only chance was to activate the engines for controlling the spacecraft's re-entry into the Earth's atmosphere. "Do it," I said. "We both knew that if this didn't work we were dead."[25]

Despite his dizziness, Armstrong somehow managed to find the appropriate control. "It still amazes me that Neil managed to do what he did," Scott recalled. "Only our intensive training and Neil's calm and cool demeanor under conditions of extreme danger managed to pull us through."[26] The desperate maneuver worked, and the spinning slowed. For the moment, the crew was out of trouble, though they still had to get back to Earth.

Back in Washington, Humphrey was just starting to speak, aware that astronauts might be dying out in space. Seamans recalls:

He said, "Now I hope we'll have good news before I finish my speech." . . . I was getting frequent information being brought to me from behind the podium. . . . Vice President Humphrey started speaking and . . . he kept lookin' at me and I'd look at him and shake my head. And then I got a message that it appeared that everything was under control. . . . So I nodded to him to make an announcement.[27]

It turned out that the problem had resulted because one of the thrusters had remained stuck in the "on" position during the docking maneuver. When *Life* approached Scott and Armstrong for a story along the lines of "Our Wild Ride in Space," the two immediately refused. "There was no way we could afford speculation about problems on the program," Scott later explained. "In no time there would have been congressmen saying, 'Gee, those guys almost died. I'm not going to vote more money for NASA, 'cos if I do we might lose our boys and I don't want that on my record.'" Scott had no doubt that the successful end to an otherwise disastrous flight had enormous implications. "If we had not recovered, who knows what would have happened to America's space program? Quite likely, it would have been shut down."

The United States would have been confronted with the nightmare loss: two astronauts dying in space, stranded while their craft endlessly orbited the Earth. The cause of the problem could not have been immediately determined, resulting in paralysing uncertainty throughout the space program. "All they would have known was that two guys had died in space," Scott reflected. "The remains of *Gemini 8* would have come down maybe nine or ten years later, badly burnt through uncontrolled re-entry. The press would have gone crazy. There would have been all these senators asking why we had been sent into space in the first place." Everything, in other words, had hung on Armstrong's ability to flip a switch while spinning madly. "I believe it is quite possible the money from Congress for NASA would have dried up," Scott concluded. "That would have been the death knell for the space program."[28]

The next flight, *Gemini 9*, was supposed to have been piloted by Elliott See and Charles Bassett, but they were killed in an airplane crash on February 28, 1966, en route to the McDonnell Douglas plant in St. Louis. The deaths had a bad effect on the program. "Gemini was still pretty fragile," Scott recalled. "We'd had launch vehicles blowing up, and trouble with the Agena target vehicles. It was a crucial time. People started worrying more, fearing we were in a run of bad luck."[29]

Into the breech stepped Gene Cernan and Tom Stafford, who embarked on June 3, 1966. Again, the aim was to dock with an Agena satellite, but this time it was discovered that the docking mechanism had not deployed completely, rendering it useless. Further disappointment followed when Cernan went on an EVA, this time with an experimental astronaut maneuvering unit (AMU). The unit did not work well, causing Cernan to be quickly overcome with fatigue, rather like White the year before. "I felt as if I was wrestling an octopus," Cernan said.[30] Worried about his safety, NASA cut the exercise short.

The docking problem was finally solved with *Gemini 10*, piloted by Young and Michael Collins. On this occasion, everything went flawlessly, including a forty-minute EVA by Collins. NASA was delighted, but a niggling worry lurked at the back of everyone's mind. Docking exercises had failed not because the maneuvers were faulty nor because of pilot error, but because equipment did not perform as planned. Switches stuck, hatches didn't open, rockets exploded. Little problems had massive consequences. Astronauts had come within seconds of dying. Everything was being done to ensure that equipment was flawless, but gremlins could not be prevented from sneaking into mechanisms. It sometimes seemed as if machinery had become so complex it had acquired a personality of its own. The only response was to repeat maneuvers until reliability seemed certain. Thus, *Gemini 11*, piloted by Pete Conrad and Richard Gordon, duplicated the success of its predecessor, and then *Gemini 12* went through it all again. The latter, close to a perfect mission, was a fitting end to Gemini. Ten manned flights had taken place in just twenty months. Huge strides had been made, and techniques essential to the lunar mission had been learned. Despite the setbacks, the way forward looked bright, and Kennedy's challenge seemed achievable.

The successful series of missions was a boon to Johnson. He had done well out of the space program. In the late 1950s, space had allowed him to project himself on to the national stage. He'd also used it to advance Texas—the Manned Spaceflight Center in Houston was his reward for services rendered. Like the consummate politician he was, Johnson understood that by attaching himself to NASA, he could bask in its reflected glow. He expected that, for the rest of the decade, the space agency would provide a steady stream of good, noncontroversial news. "Somehow the problems which yesterday seemed large and ominous and insoluble, today appear much less foreboding," he remarked after a successful Gemini mission in 1965.[31]

Beneath the benign exterior, however, there lurked a demon waiting to pounce. Fulfilling the Kennedy legacy was expensive work. A
number of programs were sacrosanct: Civil Rights, the Great Society,
space. At first, these massive ventures seemed affordable, even desirable, as Senator William Proxmire explained:

> We had a most unusual kind of economic situation. . . . We had a very
> big increase in government revenues because the economy was doing
> well. And there was a feeling that we wanted to maintain those rev-
> enues and not cut taxes. It was argued that what we should do, in
> order not to slow down the economy by running surpluses, was give
> a substantial amount back through revenue sharing. Therefore, there
> was funding available to go ahead with this exciting activity in space.[32]

It all seemed so perfect. It might have been, if not for another Kennedy
present—war in Indochina. That was one bit of "revenue sharing" too
many. Eventually, an out-of-control budget meant that LBJ had to go
cap in hand to Congress. Wilbur Mills, chairman of the Ways and Means
Committee, promised to grant the president the tax increase he requested, but only on the condition that he cut domestic spending. LBJ
would have preferred to cut military expenditure—in particular, what
he called "that bitch of a war." Civil Rights and the Great Society were
his programs, more than they ever were Kennedy's. Cutting them
would be like drowning his baby. In any case, a wave of riots in America's inner cities underlined the fact that the problems of America's
black population needed urgent attention, not to mention piles of
money. That left the space program as the most logical target for cuts.

Enthusiasm for NASA was a manifestation of socioeconomic standing. Those in steady jobs were much more likely to support the space
program than those on welfare. Blacks were less enthusiastic than
whites, high school dropouts less than college graduates. In the early
years, even though it was quite clear that rockets were very expensive,
space did not have a direct impact upon the disposable income of employed Americans. The cost seemed affordable, since it had not led directly to tax rises. Fifty cents a week was a small price to pay for all that
excitement. But for those in poverty, NASA seemed a cruel manifestation of national priorities. Going to the Moon was, it appeared, more
important than education, welfare, health, or housing. On the margins
of society, a constant refrain was heard: "If we can send a man to the

Moon, why can't our children read?" "For the poor, the Moon shot seems just another stunt," Whitney Young of the National Urban League, commented at the time. "A circus act. A marvellous trick that leaves their poverty untouched. It will cost thirty-five billion dollars to put two men on the Moon. It would take ten billion dollars to lift every poor person in this country above the official poverty standard this year. Something is wrong somewhere."[33]

Webb always argued that the lessons in large-scale management learned from going to the Moon could be applied to problems back on Earth, like curing the sick and housing the poor. According to Seamans, he "had a very definite idea about how to manage large institutions. And he really wanted to have the management of NASA as his final, great undertaking. People would say, 'How do you manage something?' They'd say, 'You do it the way Jim Webb did at NASA.'"[34] Johnson apparently felt the same, or so he claimed. "Space was the platform from which the social revolution of the 1960s was launched," he said in one of the most dubious lines in the fantasy he called his memoirs. "If we could send a man to the Moon, we knew we should be able to send a poor boy to school and to provide decent medical care for the aged."[35] But the problem wasn't management or imagination or inspiration. The problem was money. There simply wasn't enough of it to fight a war, re-create society, and embark on an adventure in space. Something had to give. As the *New York Times* astutely observed:

> The present awkward situation is part of the price for the almost im-
> pulsive way in which the Apollo program was born a half-decade ago,
> the product more of desire to buoy national morale after the Bay of
> Pigs fiasco than of rational evaluation. In 1961 the voices calling for
> sober weighing of alternatives and principles were few and largely un-
> heard; today those voices are many and powerful, and their impact is
> strengthened by the demands of a Vietnam struggle whose dimen-
> sions were unanticipated then.[36]

The job of fixing this mess fell to Charles Schultze, the White House budget director. "I was very sceptical about the worth of a manned space program," he later confessed. "I had always thought that there really weren't a lot of benefits from that—it was a great show, but not too many benefits . . . relative to the cost, because so much of the cost went to just life-support systems for the astronauts and not for science."

Johnson, however, made it clear that the lunar program was untouchable. Nothing would be allowed to get in the way of realising Kennedy's vision.

An adherent of the Wimpy school of money management, Johnson told Schultze that he would gladly pay him Tuesday for a space program today. As the latter recalled, "Lyndon Johnson would . . . explicitly say: 'A year from now I've gotta take some pain' . . . he'd fight to the last minute to try and avoid it and y'know, you never quite knew what was going to happen."[37] The solution Schultze eventually reached gave NASA what it needed to complete the lunar mission, but little more. The 1965 budget, the first over which Johnson had complete control, gave $5.25 billion to NASA, $150 million more than it had received in the previous year. After taking account of inflation, however, that was in truth a reduction. From that year forward, the allocation edged steadily downwards, reaching $3.991 billion in FY1969. The message was clear: NASA would be able to get to the Moon, but should not plan on anything big after that. Webb reacted bitterly, chastising Johnson for his failure to stand up to Congress. "There has not been a single important new space project started since you became President," he wrote on December 17, 1966. "Under the 1968 guidelines very little looking to the future can be done next year. . . . I cannot avoid the feeling that this is not in the best interests of the country."[38]

One of the reasons the space program was so costly was because hundreds of ideas were tried simultaneously, in the hope that the best one would emerge. The money wasted on plans that ended up on the scrap heap was prodigious. One such idea was "edible structure," which tried to address the problem of astronauts marooned in space or on the Moon, struggling to stay alive while a rescue mission was organized. Grumman came up with a mash that looked like dog food and could be molded into boards. Hungry astronauts could, in other words, start eating their spacecraft. The stuff tasted terrible, but then so did so many of the other instant foods dreamed up in the 1960s. As a spin-off, it seemed to have promise. When the Marines heard about the idea, they got very excited. But, as a material for building a spacecraft, it proved a dead end. It was far too heavy and, in any case, a stranded astronaut would run out of oxygen long before he starved to death.

The industries that built the rockets reaped enormous rewards for their respective areas, particularly in the South and West. Complaints

about the military industrial complex were muted because the contractors—North American Aviation, Grumman, and McDonnell Douglas—
were engaged in supposedly benign work for the space agency. At those
three firms between 75 and 100 percent of the employees were engaged
on government contracts. McNamara was not the only one to realize
that, without the space program, massive layoffs in the aerospace industry would have been necessary. The program was presented as a scientific endeavor, but it was in truth a huge public works project designed to keep Americans earning and spending.

The space program proved a special boon to the South, with its various installations in Huntsville, New Orleans, Cape Canaveral, and
Houston. Some people hoped that this would provide the economic regeneration that would inspire a social transformation—the South
would leave behind its racist ways and soar into space. But this did not
happen. The sophisticated nature of the work demanded a well-edu-
cated, highly trained workforce. As late as 1972, little more than 3 percent of scientists and engineers working for NASA were black. Granted,
there was some manual work for those lower down the social ladder,
but, when the contraction began during the Johnson years, the effect
was profound. Workers who had left the agrarian sector in order to par-
ticipate in the lunar challenge found themselves thrown on the scrap
heap.

Between 1950 and 1968 the population of Brevard County in
Florida grew from 23,000 to 239,000. Engineers, scientists, manual laborers, and assorted space groupies flocked to the area, either out of financial necessity or because of the irresistible pull of a fantasy. Jessie
Lee Kercheval moved to Cocoa in 1966, at the age of ten, when her father took a job as a teacher. The incessant development all around her,
and the perpetual talk of space (almost every child had a parent involved in some way) made her feel like she was a citizen of the future.
She lived in a new development called Lunar Heights—the name itself
suggested great optimism. The new houses had air conditioning and
mosquito control, proof that science could tame an otherwise inhospitable climate. The neighborhood was set out like a miniature solar
system, with concentric rings of virtually identical houses orbiting a
"Sun" of four houses in the center. From the streets outside her home,
where she played traditional childhood games updated to incorporate
a space theme, she watched rockets carry Gemini and Apollo into orbit.

Everything seemed possible. She grew up believing that by the time she finished high school there'd be a colony on Mars.[39]

For NASA, going to the Moon was fun. The great problem with fun things is that they sometimes seem important, even when they're not. Behind all of the esoteric reasons NASA staffers give for why the space quest should have continued lurks a selfish desire to perpetuate the good times. Glynn Lunney described it rather honestly when he recalled:

> We loved what we were able to do. We loved the opportunity to be able to do it. We loved the . . . newness and uniqueness of it, and the fact that we were able to participate in it. I loved the privilege of being in the role that I was. . . . We loved it. We loved the work, we loved the comradeship, we loved the competition, we loved the sense of doing something that was important to our fellow Americans. We were obsessed with it.[40]

Most of America was enthralled by the space quest. Few seemed to question its logic. The whole thing took on the character of an endless adventure serial with each new installment outdoing the previous one in drama and excitement. There was a goal, but no long-term plan. "Nothing was thought about what life would be like afterwards," Jan Armstrong later reflected. "Nobody had time to even think about that. The guys would come off one flight and go right on another."[41]

Lunney described the effect this frenzied activity had on a group of quite ordinary people: "I think we all carry a sense that we were involved in something that was much bigger than each of us as individuals. What we ended up doing in Apollo was a lot bigger than the sum of all of us . . . we all carry a sense of that kind of watershed event in history." But all races have a finish line. "When I was doing this work in the '60s, I never thought about what was going to come afterwards," he reflected. "I had this sense that it was going to go on forever. It's probably a failing or an idealism of youth, I guess, but you sort of think like, God, this is wonderful stuff, and you think it's going to go on forever."[42]

Nothing lasts forever. Budget cuts inevitably constrained the vision. With the main design work done, engineers and scientists jumped ship. Some 77,000 high-level engineers worked for NASA or its contractors in 1966. Two years later, the figure had fallen to 51,350.[43] The lucky ones found work elsewhere since "rocket scientist" looks good on a CV. But quite a few ended up driving taxis.

One Florida property developer tried to turn the fascination for space into an opportunity for riches. He purchased a chunk of land, subdivided it into plots, laid streets, installed lighting and plumbing, and gave his new community the name Rocket City. That was in 1966. Only six houses were ever built. Even before Neil Armstrong walked on the Moon, disgruntled workers were walking out of the Cape Kennedy complex, their severance notices in hand. The heroic days, kicked off by the magnificent men of Mercury, were already a memory. Blink, and you might have missed it.

When a rose reaches its peak of beauty, it is already in a state of decay. The same was true of NASA. Glory still lay ahead, but demise had already begun.

11

Sacrifices on the Altar of St. John

Roy Neal, the former NBC reporter who covered NASA from the early days of Mercury through the space shuttle, once revealed that the first astronauts were all required to give an "obit" interview to be played in the event they were killed in the line of duty. The standard message went: "Exploring space is a dangerous business and lives will be lost. I am not afraid to die doing what I love, nor should my death in any way cause delays in the conquest of space." Glenn, Cooper, Schirra and Shepard all performed the mandatory duty.

So too did Gus Grissom. Since the men were stating what they believed, the task was easy. During the first week of January 1967, Grissom reiterated those thoughts when he told a reporter: "We're in a risky business, and we hope if anything happens to us, it will not delay the program. The conquest of space is worth the risk of life. . . . Our God-given curiosity will force us to go there ourselves because in the final analysis only man can fully evaluate the Moon in terms understandable to other men."[1] He then went back to the preparations for the first flight of Apollo.

Despite occasional glitches, Gemini had gone well. Kennedy's goal seemed reachable, perhaps even with time to spare. The first Apollo missions, like Gemini, would test the equipment and maneuvers essential for a trip to the Moon. *Apollo 1*, with its crew of Grissom, Ed White, and Roger Chaffee, was supposed to be the first field test of the new vehicle. It would spend two weeks in space, monitor all the functions, and then test its ability to withstand reentry. Launch was scheduled for February 21, 1967.

From the start things did not go well. Development was proceeding at the same time as training, with the result that a delay in one automatically implied a delay in the other. Contractors found it difficult keeping to NASA's rigorous schedule. Apollo was so hugely complex that the astronauts spent week after week going over the systems again and again. Grissom, White, and Chaffee, along with the backup crew of

David Scott, Rusty Schweickart, and Jim McDivitt, were sent to Downey, California, to liase with North American, builder of the capsule. "The six of us spent almost every waking hour together out at Downey," Scott recalls.

> Most of the time we stayed at a motel around the corner from the plant. We pretty much lived with the spacecraft as it was being built and tested. There were even bunks at the plant so that we could sleep there; and lots of tests were conducted during the night. It was a twenty-four-hours-a-day, seven-days-a-week operation. Apollo was behind schedule.

Every venture was a pioneering one, since no one had ever designed a craft to go to the Moon. "Sometimes there were engineering paper strip sheets laid out across tables running for twenty feet. . . . No one had anticipated how many problems there would be and how much it would cost to fix them."[2]

Time was one problem, cooperation another. Hundreds of different subcontractors worked on the project, the result of Webb's desire that every state in the Union should benefit in some way from NASA largesse. But getting all these engineers to communicate with one another was hugely difficult. "They all had different cultures, used different terminology and even had different ways of making technical drawings," Scott recalls. It was left to the men at the cutting edge, those who would have to fly the craft in space, to make sure that everything worked in harmony. "It was like taking a hundred-piece orchestra, in which none of the musicians had ever played together before, and trying to get them to play a symphony, straight off, without practice. When we arrived, the Apollo orchestra was seriously out of tune. There were some beautiful instrumental solos, but they were being drowned by too much percussion."[3]

At the back of everyone's mind was Kennedy's arbitrary deadline of landing on the Moon by the end of the decade, which meant that the entire program proceeded at a pace faster than prudence dictated. The deadline, Scott felt,

> had imposed such pressure on the Apollo program that there was no time for successive redesigns of the spacecraft's hardware as the systems became more and more refined and integrated. The speed with

which Apollo was being developed also meant that a lot of the design lessons learnt during the Gemini missions were appreciated too late to be incorporated. . . . This was to contribute to a fatal flaw.[4]

"It really amazed me how quick the turnaround was between the end of Gemini and the first of the Apollo program," recalled Gene Kranz, chief of Flight Control Operations at Houston. "At times I felt like I had a foot in each program and couldn't quite figure out which way to go." It seemed strange to him that the first Apollo mission would be manned. "Always before, we had been very conservative in the development of our test programs so that we'd do what we call 'incremental flight testing,'" he explained. "We'd basically take a step at a time and very conservatively make sure that we understood what we had learned from the previous flight. We'd have relatively long gaps between missions . . . and then develop the next mission, which was a small, baby step forward." That approach was, however, rendered a luxury since time was running out. "The race to the Moon was very real to us," Kranz recalled. "We had the lunar challenge laid out in front of us. We were three years from the end of the decade, so this did not allow too much procrastination in the directions that we were taking. So we elected to fly our first spacecraft . . . in a manned fashion."[5]

Training was hampered by problems with simulators. The Command Module simulator at the Cape was particularly problematic, causing intense frustration among the astronauts. At one point, Grissom got so fed up with it that he hung a lemon inside. He also wrote a daily memo to Joe Shea, the Apollo Spacecraft program manager, detailing all the dangerous problems he had uncovered. Little, however, could be done to answer the astronauts' complaints, since, as Scott explained, "Shea . . . had to make his decisions based on scheduling and budgetary constraints. Basically we had to accept the overall hardware design of the spacecraft as it was."[6]

"An 8-hour test would routinely, in those days, take 24 hours," Kranz recollected. Experience had taught them that teething problems were a natural part of the process of getting a new spacecraft ready. In the past, everything eventually came together. This time, however, something seemed different. Confidence proved elusive. "It always seemed that every time we'd turn a corner there were things that were left undone or answers that we didn't have or we were moving down a wrong path."[7] On January 26, tests had led to fifty-six major changes in procedure. When pressed, Shea admitted during a news conference that thousands of failures in the equipment had been uncovered but that there was always the possibility that one might escape notice and lead to a disaster. That was more sagacious than he intended.

Faith in American technology and in the can-do spirit of the astronauts led to an assumption that the rocket jocks were invincible. Mercury and Gemini had lulled the public into a false sense of security. There were quite a few crises, but those merely added to the excitement of the endeavor. They seemed to suggest that the astronauts would always find a way to rescue themselves. After all, hadn't Armstrong managed to improvise a solution to his problem while his capsule was tumbling out of control? So great was the confidence that, at one point, Webb had to warn Americans that there was likely to be a tragedy, and that it would probably be witnessed live on television.[8]

On January 27, 1967, Grissom, White, and Chaffee were checking the power systems for the command module, service module, and launch systems, in order to determine whether each could operate on its own, independent of the others. The command module was prepared as if it were to be launched, with the astronauts in the capsule in their appropriate seats (or couches). They were in space suits, and the cabin was filled with pure oxygen at high pressure so as to mimic an actual launch. This was originally supposed to be an unmanned test, but the pressure of an approaching launch date (not to mention Kennedy's goal) meant that the schedule was attenuated, with a stage skipped.[9]

Due to the tight schedule and unreliable equipment, tempers were on a knife-edge. Shortly after the test started, the astronauts became aware of a sour odor in the cabin, which they reported to the technicians. This was removed by the environmental control system, but then quickly returned, whereupon it was removed a second time. Problems of this sort seemed merely annoying, not dangerous. The astronauts worried that they would inevitably lengthen the tests that day.

The mock countdown resumed, only for it to be interrupted once again, this time when problems in the communications system reduced messages between the commander and the ground staff to a garbled mess of static. By this stage the crew had been in the capsule five hours, yet had made no real progress toward accomplishing even the first objectives of that day. Then, suddenly, a fault in the electrical system produced a spark. Within ten seconds, flames were licking up from the floor on Grissom's side of the capsule. He shouted: "I've got fire in the cockpit!" White, and then Chaffee chimed: "Fire!" "Fire!"

Grissom tried to release the lever that would vent the atmosphere of pure oxygen. Meanwhile, White was desperately trying to undo the hatch. In deference to Grissom's near tragedy in his Mercury flight, the hatch had been made more secure, to avoid inadvertent opening. Those improvements turned the capsule into a death trap. The heat of the fire quickly reached 2,000 degrees Fahrenheit, but that's not what killed the astronauts. They died of asphyxiation. Out in Houston, flight director Chris Kraft was monitoring White's medical data. He noted a rapid rise in his pulse rate and then, after fourteen seconds, an abrupt halt. At the consoles around Kraft, three technicians fainted. Chaffee was found in his couch, charred virtually beyond recognition. Grissom and White were a bundle of bodies beneath the hatch, their space suits fused together.

The worst part of the accident was its inherent stupidity. Everyone knew that going into space was dangerous. Everyone had seen rockets misfiring or blowing up shortly after takeoff. Everyone had imagined the horror of an astronaut stuck in space. But this was a simple fire on a stationery rocket during a routine test. It seemed so avoidable. Surely there were ways to blow a hatch in an emergency, or to vent the volatile oxygen inside the capsule. Why hadn't anyone considered this danger? "I don't think that what happened was a case of malfeasance," Robert Seamans felt.

> There was a lot of poor workmanship but the simple fact of the matter was: we forgot one very important test and we were stupid that we forgot it. But we should have taken a boilerplate capsule and put similar kinds of equipment in it and started a fire and see what had happened. We tested all the rocket engines, we tested the stages, we tested everything up, down, and sideways and the one thing we never tested was a capsule on fire.[10]

Apollo was loaded with flammable materials. Flight plans were printed on plain paper. Suits were made of highly combustible nylon and couches were stuffed with polyester foam. Velcro was everywhere—it kept things from floating about the capsule in zero gravity. The cooling system used ethylene glycol, a highly flammable substance that produces toxic smoke. "That high-pressure pad test was a fire waiting to happen," Kraft felt. "All it needed was an ignition source. There were plenty of candidates. There were frayed and even bare wires after the command module reached the Cape, all the result of shoddy work-manship at North American."[11] Astonishing as it seems, the capsule that was supposed to take Americans to the moon would not have passed today's workplace fire safety regulations.

Kranz, who had gone home for the night, rushed back to Mission Control when he heard the news.

> I've never seen . . . a group of men, so shaken . . . the majority of the controllers were kids fresh out of college in their early twenties. Every-one had gone through this agony of listening to this crew over the 16 seconds while they . . . suffocated, . . . it was very fresh, very real and there were many of the controllers who just couldn't cope with this disaster.

On the following Monday, Kranz called together his entire team in an auditorium.

> I gave what my controllers came to know as the "tough and compe-tent" speech, and concluded the talk identifying that the problem throughout all of our preparation for *Apollo 1* was the fact that we were not tough enough; we were avoiding our responsibilities, we had not assumed the accountability we should have for what was going on during that day's test. We had the opportunity to call it all off, to say "This isn't right. Let's shut it down," and none of us did. So basically the toughness was from that day forward we would stand for doing everything right, literally being perfect and competent.

"We had become very complacent," he feels. The program had become a victim of its own success. "We all knew this was dangerous. Many of us who flew aircraft knew it was extremely dangerous, but we had sort of stopped learning."[12]

Grissom, aged 40, and Chaffee, just 31, were buried with full mili-tary honors in Arlington Cemetery, while White, 36, was laid to rest at his alma mater, West Point. The loss was keenly felt because so much emotion was invested in Apollo. NASA marketers had done their job well; by the middle of the decade most Americans were in love with the astronauts. Though the cost of the program occasionally caused disquiet, the frequency of the Gemini missions and the clear step-by-step progression had suggested that the Moon was reachable. The astronauts themselves were essential to this love affair, since Americans would never have fallen hopelessly in love with a program designed to put a robot on the Moon. Human beings might not have been essential to the concrete task of landing on the lunar surface, but they were absolutely essential to the way the program was sold to the public, and to the way it was funded.

Life tried to graft the tragedy onto the heroic story it had been peddling since 1959. The prose could not have been purpler if it had been written with plum juice. Grissom, White, and Chaffee had "bought the farm right on the pad, cooked in the silvery furnaces of their spacesuits, killed in a practice run before they could ever know the surge of their great Apollo craft driving upward to orbit. But put these astronauts high on the list of the men who really count."[13] More realistic (and worrying for NASA) was an editorial in the *Boston Globe* which alleged that the astronauts were victims of shoddy engineering and poor management, compounded by the urgency of a race. "Space exploration," the article continued, "is not the same as, say, an international boat race on the Charles. The ultimate in precautions is a first requisite. The precautions have obviously been lacking. If the lack is not downright criminal, it is little short of it."[14]

NASA's long honeymoon was over.

By 1967, President Johnson's budget problems were out of control. His sacred programs—Civil Rights, the Great Society, space—were supposedly untouchable. But the spiraling cost of the Vietnam War rendered economies essential. The Apollo fire seemed to offer a way out of the financial quagmire. Since the tragedy would inevitably place NASA seriously behind schedule, it made sense to ease up on the accelerator. "I thought there was some possibility it wouldn't make the date anyway so just slow it down a little," Charles Schultze advised LBJ. He also suggested that "it would be better to abandon this goal now in the name of competing national priorities, than to give it up unwillingly a year from now because of technical problems." But the president refused to listen. If NASA failed to realize Kennedy's vision, he did not want people to conclude that it was because he had withheld the money. "Getting to the Moon was in concrete," Schultze realized. "Getting to the Moon this decade was at least in soft concrete hardening rapidly."[15]

Wilbur Mills, chairman of the Ways and Means Committee, was, however, demanding that LBJ cut at least $600 million from NASA's budget. Schultze would have been happy to find that savings by cutting back on the manned lunar program. Had he done so, the goal of landing a man on the Moon by the end of the decade would have been abandoned. Further cuts to NASA would inevitably have followed and the lunar mission might never have been realized. Instead, Johnson put a fence around the Moon and insisted that cuts would have to come from those programs scheduled to follow the lunar landing. This was a momentous decision. Nothing would stand in the way of NASA achieving Kennedy's goal, except perhaps NASA itself. The decision also confirmed something that should have been obvious, but few stopped to contemplate. There was no such thing as a space program per se; no long-term plan to explore the heavens for the advancement of science and the good of mankind. There was instead, simply, a shallow race to get to the Moon, which, because it was once eloquently endorsed by Kennedy, had taken on the status of the Holy Grail.

Johnson decided to ignore Schultze's advice for the very simple reason that he still believed the race was important. Webb had been feeding him with scare stories that the Russians were still ahead in space and might even win the race to the Moon. Johnson remembered intimately how embarrassing it had been for Eisenhower to witness the Russians surge ahead in space. Since that time, Americans had clawed their way into the lead. The program had been sold to the American people as a life or death issue. "I hope we beat Russia to the Moon," a small boy wrote LBJ in 1968. "I know if we don't beat them they will take over the world." Children believed that line because adults had fed it to them. Adults believed it because politicians had shouted it from the rooftops. For this reason, Johnson was not about to let the Soviets back into the race. Sputnik would be tiny compared to the consequences of having the Russians win the race to the Moon. The president could not allow that to happen to America.

The fire hadn't just destroyed the lives of three astronauts, it also destroyed the confidence of NASA, an organization previously incapable of doing anything wrong. Webb had boasted openly about the "NASA system"; his mantra went: "Project management is really the art of doing what you said you would."[16] He repeatedly stressed that success was all about management, that the technology had already been developed; it was simply a case of good managers bringing science and people together. But then it all unraveled. Brilliant management had failed to prevent a stupid fire. Though Americans would get to the Moon, NASA would never be the same.

James Webb's management style was not really a style at all, but rather a collection of like-minded individuals working harmoniously on a project in which they believed. The proof of the pudding came when things started to go wrong. The system was built upon trust; Webb had the confidence that his engineers and technicians would tell him things that he needed to know, especially since he was not an engineer and could not be expected to find out on his own. The fire destroyed that trust, but more importantly it revealed that trust by itself was not enough to ensure the smooth running of a huge operation like going to the Moon. An orderly system of monitoring had never been established.

The biggest problem was North American Aviation. NASA engineers felt increasingly uneasy about the quality of workmanship. Long before the fire, they decided to commission what was called a "Tiger Team" to investigate. The team, headed by General Sam Phillips, the program director of Apollo, filed a highly critical report in late 1965. It alleged that too many shortcuts were being taken, that North American was in way over its head. Components were riddled with flaws, yet this was an endeavor in which the slightest error could cause major catastrophe. The report complained:

> Poor workmanship is evidenced by the continual high rates of rejection and . . . actions which result in rework that would not be necessary if the workmanship had been good. . . . Today, after 4 1/2 years and a little more than a year before first flight, there are still significant technical problems and unknowns affecting the stage. Manufacture is at least 5 months behind schedule. . . . NAA performance . . . is characterized by continued failure to meet committed schedule dates with required technical performance and within costs. There is no evidence of current improvement in NAA's management of these programs of the magnitude required to give confidence that NAA performance will improve at the rate required to meet established Apollo program objectives.[17]

The need for haste had caused corners to be cut. Engineers at the Marshall Flight Center regularly took apart rocket components that North American supplied, in order to make sure they weren't dangerous. Webb objected to this practice because he felt that it was the contractor's responsibility to make sure that a product was flawless. Shortly after directing von Braun's team to desist, one of the Marshall engineers showed up at Webb's office with a rag he'd found inside the bowels of a component supplied by North American. According to George Mueller, the problems were "so comprehensive that it would have been very difficult . . . to explain why we were continuing to go on with [North American], when they were doing such a terrible job."[18]

NASA blamed North American, but many contractors blamed NASA. The buck was supposed to stop with them; they were the ones doing the final inspections and, eventually, approving a piece of machinery for flight. Many of those involved in the construction of the lunar package feared that speed was taking precedence over safety. Mel Friedman, a middle-level quality control inspector, recalled a deep anxiety:

> From my end, I wasn't supposed to accept anything that did not meet the criteria one hundred percent. If, during a test, a piece of equipment was supposed to give a reading of ten volts, plus or minus one volt, that was it: nine to eleven volts. Nothing else. I was surprised to learn that this rule was being disregarded almost all the time. . . . If it didn't come right up to specification, NASA would let it pass if our constraints were going to delay the program in any way. . . . My supervisor . . . seemed determined to meet Kennedy's deadline, and in my opinion a lot of risks were taken because of that. . . .
>
> It troubles me to this day, that the kind of quality control we were fighting for early in the program was not demanded by NASA until later in the program, until *after* the first men landed on the Moon. You think it would have been the other way around.

Lynn Radcliffe also felt the pressure of deadlines. In a letter to Joe Gavin, the LM program manager, written toward the end of 1966, Radcliffe reacted angrily to news that the White Sands test facility (of which he was manager) would be shut down in September 1967:

> Because of the schedule, we should get to the Moon. Kennedy said we will do this in the decade of the 60s. Very well, then, we're going to do

it for John. But all that we have done here to date, and all that we will have done by September of 1967, will be to prove that the LM's propulsion systems . . . can work. We will not have shown that they will work, first time, every time. We have proven only the concept . . . We need more engineers assigned down here, not fewer. We must keep right on going. As we find out things that need to be checked, we want to have a team here that can do it. They say we are in a race with the decade. This is dangerous. If we do not change our attitude about schedule and start worrying more about quality, we're going to lose some astronauts.

Ironically, the people who seemed to care the least were the astronauts themselves. All the worrying was done for them. "All the astronauts I met, Neil Armstrong and all of them, were different from the rest of us," Milt Radimer, one of the senior engineers at Grumman, thought. "[The astronauts] believed in everything. They always believed they were going to do it. There was never any hesitation in their minds about whether the ship was going to work."[19]

In the past, when something went wrong, NASA automatically commissioned an internal inquiry, on the grounds that those most deeply involved in the project were best placed to investigate. That was the way the military did things and NASA saw no reason to do it differently. But with a tragedy this big and this public, an internal inquiry looked suspiciously like a cover-up. "There must be some doubts as to the impartiality of the special board now investigating the fatal fire," the *Wall Street Journal* commented. "Although the panel does include experts from outside the space agency, it is heavily weighted with NASA-affiliated personnel. Perhaps more confidence would be engendered if NASA were not, in effect, investigating its own tragic accident."[20] Webb ignored those complaints, because he feared that an outside investigation would be used by NASA's enemies to kill the lunar mission, if only through delay.

Congress, however, refused to cooperate. One month after the fire, the Senate Aeronautical and Space Sciences Committee began formal hearings into the accident. Walter Mondale, a junior member of that committee, was already aware of management problems within NASA. "I was starting to pick up comments from people in and around the program along two lines. They were having management problems. There were risks that were not being fully dealt with. . . . That they were having trouble with the private contracts, that they were in a hurry and might be taking some chances."[21] Mondale had gained access to a huge stack of photos from NASA's internal investigation. One of them was a close-up of a wrench that a careless technician had left deep inside a tangle of wiring.

An anonymous source from within North American leaked the existence of the Phillips report to Mondale. From his perspective, it appeared that NASA was trying to bury evidence that it knew of a problem with the contractor long before the fire. At the hearings, he dropped a bombshell during his questioning of Webb:

> I have been told, and I would like to have this set straight if I am wrong, that there was a report prepared for NASA by General Phillips, completed in mid or late 1965 which very seriously criticized the operation of the Apollo program for multi-million dollar overruns and for what was regarded as very serious inadequacies in terms of quality control. This report . . . was so critical that it recommended the possibility of searching for a second source, and as I am told, recommended Douglas Aircraft. . . . Would you comment on that? Is there a Phillips report?[22]

Webb felt like he had been blindsided by a truck. He claimed that he had never heard of the Phillips report. Mondale thought he was lying. It's quite possible that he was actually telling the truth, but in these circumstances, not knowing about the report was probably more damning than knowing about it and trying to bury it. Seamans, who did know about it, finds it mysterious that he somehow neglected to tell Webb. "For the life of me, I can't believe that I didn't keep him informed. But he claimed after the fire that no one had ever told him about these problems. . . . That's one of the things I feel badly about: that . . . he felt that he'd been kept out of the loop."[23]

Mueller certainly knew about the report, but never expected it to come out. "I guess I was surprised . . . when this Phillips Report began to surface, because . . . I had thought it was carefully buried and destroyed."[24] When questioned by Mondale, he made light of the issue by claiming that it was merely a routine evaluation of a contractor. Mondale then asked whether Phillips had recommended looking for a second contractor to replace North American. Mueller replied that he had no knowledge of such a recommendation. That was a lie, since Mueller had referred directly to the report in a December 19, 1965, letter to Lee Atwood, president of North American. "I consider the present situation to be intolerable," he told Atwood, "and can only conclude that drastic action is in the best national interest."[25] Already deep in a hole, Mueller should have stopped digging. He should have realized that, once Mondale got wind of the report, it was inevitable that a copy would surface. When it did surface, it merely confirmed that NASA had something to hide. On the following day, one newspaper remarked that NASA stood for "Never A Straight Answer."[26]

NASA, for so long the darling of public agencies, was neither accustomed to, nor prepared for, having its judgment questioned. When Seamans took the stand, he told Congress to mind its own business:

> For a relationship to be effective between Government and contractor, there must be mutual confidence and if on every occasion that there is either a minor or a major review this is going to be exposed in all of its detail, it will soon erode the confidence that is so necessary. . . . The release of this report would seriously impair our ability to prosecute work in the future.[27]

Mondale was surprised at how NASA considered itself above reproof. "They didn't want to disclose it. They didn't want a public discussion of the problems that the space program was into."[28] Meanwhile, Webb turned on Seamans for revealing too much:

> I remember it so clearly . . . Mr. Webb nailing me on the floor there of the hearing room saying: "I want you to come with me" and as soon as we got in the car . . . it had a window in it that you could crank up between the driver and the back seat and he cranked that back window up and he really turned on me, he said, "There's really no excuse for you volunteering information at that hearing." He said, "We're dealing here with matters that can result in millions and millions of dollars of lawsuit and this is not like the kind of friendly hearing that you're used to."*

* In 1972, North American paid a total of $650,000 to the widows of the three astronauts who were killed.

According to Seamans, it was only after he and Webb had returned to their offices that he told Webb about the existence of the Phillips Report. "Oh my gosh!" Webb reacted, "I have lied to Congress!"[29]

Webb then went to see Mondale and suggested that the senator had a duty to discuss the matter with him privately, instead of raising it in a public forum. "I said, 'You know, Mr. Webb, I don't agree with ya. I'm a United States Senator, this is public business. And I think I did exactly the right thing and I'm not gonna get in a position where I have to ask permission to ask questions. I'll reserve that right, I'm a Senator.'"[30] Despite all these embarrassing revelations, however, NASA emerged from the inquiry surprisingly unscathed. The agency was basically told to get on with the task of fulfilling Kennedy's challenge. Most Americans, as long as they didn't think too hard about the cost, still wanted NASA to take them to the Moon. The Senate, undoubtedly aware of the public mood, was hesitant to censure the agency. Its final judgment said that the Phillips report "had no effect on the accident, did not lead to the accident, and [was] not related to the accident." NASA's handling of the affair did, however, dent its reputation, not to mention its self-belief. In the space of a year it went from being an almost unbelievably perfect organization to a rather ordinary one in which recrimination, back stabbing, jealousy, vindictiveness, and subterfuge seemed commonplace.

Webb's perfect organization had begun to crumble. In particular, he felt let down by Seamans. "Our relationship had been exemplary for about 6 years," Seamans commented. "He'd been just a wonderful boss. . . . And all of a sudden I felt as though he had me sort of under the gun."[31] Webb's sense of betrayal soon began to filter down through the organization. "Jim began to lose confidence in Bob, and of course he lost confidence in me, and he lost confidence in all of the management team," Mueller recalled.[32] Before long, Seamans decided it was time to go, citing the fact that he was "tired of trying to keep the Apollo program on track in the face of . . . disputes with . . . Jim Webb."[33] He jumped before he was pushed. Others were not so quick. Shea, directly responsible for the Command Module, was "reassigned." Further down the line, there were additional casualties. At North American, a heavy cull took place.

Eventually, Webb would also leave. Tired of dealing with the yearly cuts to his budget, he had attempted to embarrass the Johnson administration by openly suggesting that the United States might actually lose the race to the Moon. On a number of occasions, he had threatened to resign unless Johnson restored his budget to the 1965 level. Though Johnson still had enormous respect for Webb, he was getting tired of his whining. On September 16, 1968, Webb told Johnson that he was thinking of resigning in the near future. "I've been thinking along the same lines," the president replied. "Let's step outside and tell the Press that you are leaving, effective immediately!"[34] In his place came Thomas Paine, who had earlier replaced Seamans.

In hindsight, everyone blamed the accident on the reliance upon a pure oxygen environment, which turned a small spark into a conflagration. Alexei Leonov felt that the accident could so easily have been avoided had the Americans been willing to learn from Soviet mistakes:

> I was . . . very angry at how stubborn the American engineers were in continuing to use a pure oxygen environment in the spacecraft. I couldn't understand why they had not switched to the system we adopted after the death of Valentin Bondarenko:* re-generating oxygen during a flight.
>
> The Americans must have known of the tragedy that had befallen Bondarenko in a pure oxygen environment. He had been given a big funeral, and the American intelligence services would not have been doing their job properly if they had not informed NASA about what had happened.[35]

The actual cause of the *Apollo 1* fire was unimportant; what mattered was that any fire would have been fatal. "Hindsight is wonderful," Max Faget reflected.

> We had the same atmosphere in Mercury and Gemini as we had in Apollo. They never had any fires. But, you see, after I started thinking about it, kicking myself for being so stupid, I realized that the difference between Mercury and Apollo was that one Apollo experience was probably equivalent to maybe 20 or 30 Mercurys, simply because there's so much more volume in Apollo and there's so much more stuff in Apollo . . . It only takes a teeny bit of stuff, with some teeny bit of flammable material to ruin the whole thing . . . sooner or later it would

* Bondarenko had died on March 23, 1961, as a result of a fire in an isolation chamber.

have happened in one of the Apollo flights. It just happened to happen on this one.

Faget accepted that Apollo was deeply flawed. "There were other things that were not good in that spacecraft when you looked at it. The wiring was not too well done. It was not neatly done. There were too many fixes and so forth." The spacecraft needed a complete redesign, but the deadline was paramount. "Where they found something wrong with the electrical service, they'd tape over or jump over it. A lot of circuits were added later on that weren't anticipated, which made for junky wire bundles." Faget did not, however, believe that the tragedy had been caused because everyone was in a hurry. "No. I didn't think so. If they'd said, 'Relax, take another three or four months,' we'd still have probably flown the same spacecraft, still would probably have run the same tests, still probably had the same goddamned fire. . . . It was just a relaxed attitude towards fire, which was not called for."[36] Chris Kraft did not agree. "We got in too much of a God-damned hurry," he later admitted. "We were willing to put up with a lot of poor hardware and poor preparation in order to try to get on with the job, and a lot of us knew we were doing that."[37]

Three dead astronauts was big news, but a dead technician or engineer struggled to make the back pages of his local newspaper. During the memorial services for the *Apollo 1* astronauts, two men conducting medical experiments for NASA in a similar oxygen-rich environment were incinerated when a spark caused a conflagration. No one remembers their names. Another worker was killed when his head was crushed after getting hit by the spring-loaded lunar module leg.

Others died quietly, smothered by stress. "The pressure is excruciating—all the managers work under the subconscious strain of something going wrong," George Skurla, director of Grumman operations at Kennedy Space Center confessed in 1969.[38] "Throughout the month of June 1967, we were trying to make sense of the whole mess—and it was a mess at that point," Tom Kelly, who ran the LM assembly and test operation, recalled. "We were leaning very heavily on some of our engineers and test conductors and test directors. I mean, we were really pushing them. We were giving them deadlines and we wouldn't take any excuses. . . . Well, one day one of the fellows we were leaning on very hard collapsed right on the job." Kelly looked back with deep regret at the way nameless, faceless technicians were sacrificed at the altar of St. John. "You wouldn't believe the problems some of these men had hanging over their heads. There were drinking problems, families deteriorating, depression bordering on suicide. We hadn't known anything about this. We were just cracking the whips over their backs."

Radcliffe recalled telling Kelly: "What the hell are we doing here? We don't know anything about these people. We don't know what their physical condition is. We don't know what personal pressures they are under. And still we're pushing them and pushing them, and sometimes they break." He especially regretted the way ordinary engineers were rapidly elevated to positions where they were in charge of perhaps fifty people, without being given any training in personnel management. "We haven't given them any help," he remarked. "All we are doing is beating on them, watching them grow prematurely old before our very eyes from week to week. What the hell is the matter with us?"[39]

According to Skurla, anyone who rocked the boat was "summarily thrown off." Yet the problem wasn't just that of managers being cruel to men lower down the ladder. A lot of the wounds were self-inflicted. The Moon quest bred an obsession that could be deadly. "Apollo left a lot of human wreckage in its wake in terms of broken families, divorces, and busted professional careers," he added. "Cocoa Beach had the highest divorce rate in the country. The guys were so fascinated, especially the younger ones—so totally committed—it was an intoxicating thing." He remembered visiting a colleague seriously ill in a hospital bed whose only worry seemed to be the fact that he was falling behind schedule with his work. He died before he could get back on the treadmill. Another colleague went to work on a Saturday instead of spending it with his son, who was shipping off to Vietnam the next day. Skurla himself was bedridden for two weeks after collapsing from exhaustion.[40]

Wives of astronauts suffered terribly since they had to deal with the corrosive anxiety of flights, the prolonged absence of their husbands, and (often) his weakness for adoring female groupies. All the while they had to play the happy, perfect spouse for the benefit of the media. They were like single parents who had constantly to pretend that they were happily married and living normal lives. All that America wanted to hear was that they were "proud, pleased, and happy." "We didn't burden [our husbands] with trivial things," Jan Armstrong recalled. "We had the pressure on us to make do. All the gals were busy raising their kids. The guys blew in and out." Mueller used to hold teas for the wives, the purpose of which was to keep their spirits up. Armstrong found the sessions condescending and quickly grew annoyed with Mueller's repeated refrain: "Now, girls, you keep up the good work. Keep a stiff upper lip."[41]

Bill Voorhest admitted that he missed Sunday dinner with his family for two years running. Joe Kingfield, another Grumman employee, confessed: "It took an awful long time for wives to realize that they had lost their husbands until we got on the Moon. It was hard too, on the kids. . . . There were things that you [should] be doing normally with your children while they were growing up. And I was missing those things."[42] After the Moon landing, he turned around and noticed that his kids were grown.

12

Merry Christmas from the Moon

On May 6, 1967, at a dinner to celebrate the sixth anniversary of the first American in space, Alan Shepard signaled that it was time to resume the race:

> Much has . . . been said about the cause and effects of the fire. In this case, perhaps too much. . . . All of us here tonight jointly share the responsibilities for the human frailties which are now so apparent—and for the insidious combination of materials and equipment which was so devastating in their behavior. . . .
>
> The time for recrimination is over. We have digested enough historical evidence. There is much to be done. Morale is high. Vision is still clear.
>
> And I say, let's get on with the job.[1]

The problem was not quite as simple as Shepard suggested. The fire put the space program seriously behind schedule. "The whole capsule had to be rewired," George Mueller remembered. "And of course that took time and energy and effort. Just simply getting everybody going in the same direction at the same time was a major challenge."[2] "We weren't just looking for fire hazards," Joe Gavin recalled. "We went back and looked at every single system."[3] Rebuilding the capsule was perhaps the easiest part. Rebuilding morale was much more difficult. "There was a whole year where nobody was willing to take any risk whatsoever," Mueller reflected. "I had to take the lead in convincing people that it was safe to fly and that we really couldn't afford not to take some risks, that there wasn't any way to fly any of these things without risk."[4]

In the past, NASA had employed "incremental testing"—what Gene Kranz called "small, baby steps."[5] If going to the Moon was like climbing a ladder, each rung was meticulously tested before being used to advance to the next. But, after the fire and the inquiry, there wasn't time for that. Mueller had earlier proposed "all-up testing," or running up the ladder without bothering to test the rungs along the way. The idea, which once seemed madness, now seemed the only option. The Moon package—the Saturn V rocket, with all three stages, and the Apollo capsule on top—would be tested as a unit. If it worked, it meant that everything was fine, and tests which previously would have taken months, or years, could be achieved with one mission. But if it didn't work, and, say, the second stage misfired, no one would have any idea whether the apparatus above that stage was sound. The accelerated approach would eventually mean that NASA more than made up for the delay. "There was enough wrong with the spacecraft that, without the fire, we might not have made Kennedy's deadline at all," Kranz thought. "We'd have flown, found problems, taken months to fix them, flown again, found more problems, taken more months. . . . We might not have landed on the Moon until 1970 or '71."[6] Robert Seamans agreed: "Tragic as it was, if we had not had the accident then, we would not have gone to the Moon in the decade."[7]

On April 4, 1968, an unmanned *Apollo 6* rose from the pad at Cape Kennedy on a voyage of woe. Everything seemed to go wrong. The capsule spent a short ten hours in space before plunging into the Pacific. During its absence, James Earl Ray gunned down Martin Luther King Jr. in Memphis, thus insuring that NASA's travails were buried on the back pages of the newspapers. Jules Bergman, a reporter for ABC, nevertheless concluded:

> The troubles with Apollo 6 almost certainly mean another delay in America's lunar landing plan. But even if this flight had been a full success, the space agency is still in deep basic trouble. It is searching for the future, and it hasn't found it yet. Nor has it convinced Congress that it has a real future beyond the Moon. NASA now faces new budget cuts of up to half a billion dollars as part of the austerity drive. The great goals in conquering space are still just as great, but Congress feels there are greater, more immediate goals to be conquered here on Earth: rebuilding our cities, solving our transportation problems, exploring the oceans. And even an end to the Vietnam War won't solve NASA's problems. The faith of Congress and the space agency's credibility remain shaken.[8]

Reginald Turnill, who covered the space program for the BBC, noticed a huge change. Every aspect of the program now seemed strictly regulated, as if bureaucracy was the best answer to disaster. "That rigid dis-cipline illustrates the difference here at the Cape now, compared with ten years ago when I started coming," he reported at the time. "Then everybody went about in the heat in shorts and open shirts, felt like a pioneer and was happy. Now, everybody wears suits—ties, jackets, the lot—and of course worries."[9]

Thomas Paine was appointed deputy administrator of NASA on January 31, 1968. It seemed to him that the press was actively looking for disaster, since success no longer seemed newsworthy. "The air of an-imosity and the frank feeling I got from the reporters was that the best story they could possibly get would be to have the damn thing explode on the pad . . . in which case they would have credit lines on the front page, whereas if the damn thing went up, well there wasn't much of a story in that." Though the agency still had its journalist friends, the days when it could depend on positive reporting were over. Some jour-nalists, Paine remarked bitterly, "knew we didn't know what the hell we were doing and that we were a bunch of bums and that it was all a great façade really."[10]

Public support for the lunar program also suffered. A Harris poll taken in July 1967 found that 46 percent opposed the aim of landing a man on the Moon, while 43 percent favored it. Rather predictably, sup-port was highest among young men under 35 in the upper income bracket. As in previous polls, support fell drastically when pollsters re-ferred to the costs of the mission. When asked whether space explo-ration was worth the investment of $4 billion a year, only 34 percent agreed, with 54 percent disagreeing. While those figures were worrying enough, much more revealing were answers to the question: "If the Russians were not in space, and we were the only ones exploring space, would you favor or oppose continuing our space program at the pres-ent rate." Opposition outnumbered support at a level of 60 to 30 per-cent.[11] In other words, NASA had not managed to convince the Ameri-can people that the space program had any justification beyond that of beating the Russians.

Apollo 7, the first manned mission, was crewed by Wally Schirra, Walter Cunningham, and Donn Eisele. Launched on October 11, 1968, it stayed aloft for nearly eleven days and was the first full test of the com-bined command and service modules in orbit. For the public, it was a triumphant return, especially coming as it did after a year marred by the Tet Offensive, the assassinations of Robert Kennedy and King, race riots, the Chicago convention, and widespread campus unrest. But, while the public delighted in seeing NASA again doing what it did best, the flight was not without trouble. Schirra developed a cold, which he quickly passed on to the rest of the crew. Sneezing in zero gravity is not nice. But much worse is the fact that the absence of gravity means that the nasal passages don't drain without constant blowing of the nose. That's uncomfortable, especially when the flight controllers were insisting that the astronauts had to keep their helmets on. To make matters worse, the food was terrible, even by NASA standards. A few days into the flight, the crew started arguing with each other over the more desirable meals left in the store. Dehydrated food produces gas bubbles when reconstituted. In zero gravity, these do not escape to the surface, but are instead ingested, causing severe gas pains. The only solution was to fart, yet farting in a space capsule was not a good idea when three astronauts were desperately trying to remain civil to each other.

Schirra, who had announced beforehand that this would be his last flight, came close to open mutiny, rebelling at the intrusive behavior of Mission Control. He objected to the addition of tests that had not been part of the original flight plan and seemed an excessive demand on an already overburdened crew. "I have had it up to here today," he said on one occasion. "From now on I am going to be an onboard flight director. . . . We are not going to accept any new games . . . or do . . . some crazy test we never heard of before."[12] He didn't like the way the PR people had imposed themselves on the mission plan, adding media stunts for the simple reason that NASA wanted to be on television with good news.* "I had fun with Mercury. I had fun with Gemini," he later explained in his defense. Then things changed and it was no longer fun. "I lose a buddy, my next-door-neighbor, Gus, one of our seven; I lose two other guys I thought the world of. . . . I was assigned a mission where I had to put it back on track like Humpty-Dumpty."[13] By the end of the mission, communication between the spacecraft and Houston oozed contempt. This sort of rebellion would not, however, be tolerated. The Schirra problem solved itself with his preflight decision to retire. But Eisele and Cunningham were never again given the opportunity to fly. "I didn't want to see either of them in a command position," Kraft explained.[14] They were quietly shoved aside while the American

* A short time later, Schirra joined the CBS news team, providing expert opinion as counterpoint to Walter Cronkite's commentary.

public was encouraged to believe that the space program was back on track.

On the morning of December 8, 1968, Pavel Belyayev was waiting in *Zond 7* on the launch pad at Tyuratam in Kazakhstan, while technicians made their way through a long countdown. The itinerary for his trip was to go to the Moon, circle it, and come back. But, four hours before the scheduled launch, he was removed from the capsule because readings showed problems with the first stage booster. A long inspection followed and the rocket was passed fit, but it was decided that the trip to the Moon would be unmanned. The rocket made it to an elevation of about 27 miles. Then it blew up. NASA enjoyed an early Christmas at the expense of the Soviets.

The next Apollo flight was supposed to test the command and lunar modules in Earth orbit. But because the construction of the LM had been delayed, that sort of profile could not be achieved on schedule. That meant another test of the command and service modules in Earth orbit. "Kraft and many of us felt that this was sort of a nonsensical mission," Kranz recalled. "It was just too darned conservative for the tight program we were flying, and it was really going to delay us getting to the Moon. So he proposed that we take this . . . mission, which would go to 4,000 miles, only instead . . . take it 250,000 miles to the Moon."[15] As Wernher von Braun quipped, "Once you decide to man a Saturn V, it does not matter how far you go."[16] But that was typical von Braun bluster. Deke Slayton and Alan Shepard later claimed that the mission was "the single greatest gamble in space flight then, and since." At the time the decision was made, the systems were not remotely ready. "Shit, we didn't even have the software to fly Apollo in Earth orbit, much less to the Moon," Slayton claimed.[17]

When technicians crunched the numbers in the computers, four possible launch windows came up. One of these had the crew circling the Moon on Christmas Eve, which made the decision rather easy. The crew of *Apollo 8* (Frank Borman, Jim Lovell, and William Anders) would take the command and service modules into orbit around the Moon, in the process testing the worthiness of the craft for the lunar voyage and practicing the procedures essential for future lunar missions. The improvised plan would allow tests of the communications system at great distances and also permit close-up photography of potential landing sites. More importantly, the American people would be given the Moon for Christmas. While the launch date was more coincidence than conspiracy, once it became clear that the Apollo astronauts would be in the sky at the same time as Santa, NASA made the most of its golden opportunity.

While orbiting the Moon on Christmas Eve, the astronauts conducted their own religious service to an estimated television audience of 500 million people. Anders began: "For all the people on Earth, the crew of Apollo Eight has a message we would like to send you." He began reading the first four verses of the Book of Genesis. Lovell took over and read another four verses, whereupon Borman finished the reading. He ended: "And from the crew of Apollo Eight, we close with good night, good luck, a Merry Christmas, and God bless all of you— all of you on the good Earth."[18] Kranz recalls listening to the broadcast from Mission Control:

> I think that was probably the most magical Christmas Eve I've ever experienced in my life . . . I mean, you can listen to Borman, Lovell, and Anders reading from the Book of Genesis today, but it's nothing like it was that Christmas. It was literally magic. It made you prickly. You could feel the hairs on your arms rising, and the emotion was just unbelievable.[19]

Beyond the message of goodwill, the broadcast gave a sense of transcendence to the lunar mission. It encouraged the impression that the astronauts were not simply three lonely men orbiting the Moon in an aluminium can, but true explorers of the heavens. They were, it seemed, pushing out the boundaries of human experience but doing so in a world created by God. More importantly, they were sending the message that the lunar mission was not about machines but about Man, a message which, however inaccurate, the public wanted to hear. The space race had always been more about emotion than reason, even though the chariots themselves were products of highly refined science. The service from space fit in perfectly with the emotional challenge. For most people, *Apollo 8* was a bright achievement at the end of a dismal year. One woman wrote to NASA: "Thank you for saving 1968."[20] *Time* made the three astronauts its Men of the Year. When asked on his return if the Moon was made of green cheese, Anders replied. "No, it's made of *American* cheese."[21]

After the service, the astronauts turned to their Christmas dinner, expecting the same freeze-dried fare they'd been eating since their first meal in space. But NASA had packed a surprise. They got real turkey, stuffing, cranberry sauce, and the works, in vacuum-packed containers courtesy of the U.S. Army. At that very moment, American soldiers in the jungles of Vietnam were eating exactly the same meal.

"We don't know if you can see this on the TV screen," Anders reported from lunar orbit, "but the Moon is nothing but a milky waste. Completely void." Borman then chipped in with his assessment: "It's a – a vast, lonely, forbidding expanse of nothing . . . it certainly would not appear to be a very inviting place to live or work. It makes you realize what you have back there on Earth. The Earth from here is a grand relief."[22] At a stroke, the romance of the Moon was destroyed, or at least it should have been. *Apollo 8* revealed what everyone already knew, but few were prepared to admit: that there was nothing very nice about the Moon—except for the fact that it was far away. C. S. Lewis predicted as much when he wrote in 1963: "The immemorial Moon—the Moon of the myths, the poets, the lovers—will have been taken from us forever. Part of our mind, a huge mass of our emotional wealth, will have gone. Artemis, Diana, the silver planet belonged in that fashion to all of humanity: he who first reaches it steals something from us all."[23] The crew of *Apollo 8* confirmed something Lagari Hasan Celebi discovered back in 1623: space is most important for what it means back on Earth.

"The vast loneliness of the Moon up here is awe inspiring," said Lovell during another live broadcast. That is what space buffs expected, even craved. But then came the punch line: "It makes you realize just what you have back there on Earth."[24] The dreamers didn't want to hear that. They wanted worlds more exciting, more beautiful, more bountiful than boring old Earth. It pained them to hear that ten years and $24 billion had led to the discovery that Earth was a rather nice place. But that was indeed the message behind the spectacular pictures of earthrise taken from the Moon. Environmentalists loved the photos even if they condemned the wasteful industry that had produced them. The photos became icons of the environmental movement, subsequently used on countless posters encouraging people to recycle waste, clean up the rivers, and use public transport. One of the more bizarre manifestations of the image came when ecologists began to speak of "Spaceship Earth."

Apollo 8 qualified as the most expensive trip to a viewpoint ever undertaken, and underlined the futility of the space adventure. Nevertheless, back on Earth, the pictures sent writers searching for new stocks of superlatives. President Johnson was so proud of the earthrise photo that he sent copies to all the world's leaders, including Ho Chi Minh. The U.S. Postal Service came out with a commemorative stamp showing the image of earthrise, with the message "In the beginning God. . . ." The little blue, white, and brown pearl encouraged the assumption that the world should be seen as one place, instead of as a collection of antagonistic nations. "If there is an ultimate truth to be learned from this historic flight," Johnson remarked during a moment when his hyperbole control mechanism was malfunctioning, "it may be this: There are few social or scientific or political problems which cannot be solved by men, if they truly want to solve them together."[25]

Everyone seemed to want to turn *Apollo 8* into a defining moment; everyone seemed to think that the event would render the Earth itself somehow different. Big events often cause rational commentators to assume that billions of people will somehow, overnight, decide to live their lives differently.* "This is the last day of the old world," Arthur C. Clarke remarked.[26] "The voyage of *Apollo 8* . . . did more than any single event last year to restore man's faith in himself," Thomas O'Toole commented in the *Washington Post*. "It could be that when man walks the Moon for the first time it will be felt round the world as such a triumph of the human heart that its beat shall go on for a million years."[27] The poet Archibald Macleish was similarly intoxicated by Moon juice: "To see the Earth as it truly is, small and blue and beautiful in that eternal silence where it floats, is to see ourselves as riders on the Earth together, brothers on that bright loveliness in the eternal cold—brothers who know now that they are truly brothers."[28]

Meanwhile, down in the jungles of Vietnam those brothers continued to murder one another with ever increasing ferocity and efficiency. Much as we would like to think that the planets affect our behavior, this was one instance when they clearly did not.

For NASA, *Apollo 8* provided valuable confirmation that the package which would take Americans to the Moon actually worked. *Apollo 9* then took on the original profile of *Apollo 8*, except for the fact that, given the earlier mission's success, there seemed little point in testing the lunar module in Earth orbit. The crew of James McDivitt, David Scott, and Rusty Schweickart therefore went to the Moon. After the command

---

* As was seen in the immediate aftermath of 9/11.

module separated from the spent rocket, the crew turned it around and then docked with the lunar module, which was still enclosed in Saturn's final stage. They then pulled away and headed for the Moon.

Once in the Moon's orbit, McDivitt and Schweickart climbed into the lunar module, separated it from the command module, and flew it for the first time. They practiced maneuvering with the descent engines and then completed the test by firing the ascent engines, leaving the descent stage behind before docking with the command module. Echoing Lovell, Schweickart remarked on his view of Earth, calling it "that little blue and white thing . . . all of history and music and poetry and art and death and life and love, tears, joy, games."[29]

Over in the Soviet Union, the mission had a profound effect upon Alexei Leonov, who at one time entertained thoughts of being the first man on the Moon. "I knew we were not going to beat them. Although no firm dates had been announced, we believed there would be one more Apollo mission before the ultimate goal of a lunar landing was attempted. We would never be able to accomplish the lunar missions we had planned on schedule."[30] The race was over.

The *Apollo 9* crew attended a celebration dinner at the White House after the mission. President Richard Nixon had been in office for only two months, and his real feelings about the space program remained unclear. In attendance was a Democratic senator[*] who was an outspoken opponent of the space program. A polite but determined debate ensued between the astronauts and the senator. Scott recalled: "The senator was bright, articulate and uncompromising. 'You'll never make a lunar landing by the end of the decade,' he insisted. 'We've wasted all this money. What do you think you're really going to get out of it? Why spend money on exploration when you can spend money on the homeless?'" All the while, Nixon just listened. In retrospect, Scott decided that Nixon had carefully engineered the argument in order to get the astronauts to provide concrete justification for what they were doing. The astronauts gave the standard spaceman's platitudes about spin-off technologies, the benefits to science, and the need to explore. But, as Scott later reflected,

the unspoken political undercurrent to our discussion was the importance of the space program in winning the Cold War. I did not say it di-

---

[*] Scott did not reveal his name.

rectly to the senator grilling us, but underlying my thinking were very fundamental questions: "Do you want us to win this race? Do you want to live in a free society? Or do you want to live under communism?"

The bottom line was our profound belief that we had to demonstrate democracy was a better system under which to live.

Though Nixon never gave his opinion, it seemed to Scott that the president was "pretty enthusiastic about the program."[31] Time would reveal otherwise.

One final dress rehearsal remained. The crew of *Apollo 10* (John Young, Gene Cernan, and Tom Stafford) were assigned the task of running through an entire lunar mission, stopping just short of actually landing on the Moon. They left on May 18, 1969, and executed an otherwise flawless mission, but for a single hair-raising eight seconds when the ascent stage of the lunar module was thrown into a spin due to an inadvertent setting of a switch in the wrong direction. That event demonstrated that human error would always be a factor if the crew was human. The mission itself revealed that the equipment was now ready, and that, if the passengers could perform properly, a landing on the Moon was indeed possible.

After the long delay caused by the *Apollo 1* fire, Americans had gone from a broken space program to one poised to land on the Moon, all in the space of seven quick months. Worried observers feared that everything had been done too quickly, that NASA seemed reckless. But the planners in Houston took a different view. They knew that they had the package to get to the Moon; all that was needed was for it to be tested in space. Once tested, there was no point in delaying the real thing. As George Mueller explained, the reason for flying so few test missions was the simple fact that "it got you to the Moon quicker." In this sense, it was not just a case of meeting Kennedy's deadline. "I particularly wanted to make sure that we kept the enthusiasm of the people up as we were going and to reduce the risks to an absolute minimum . . . before we got to the Moon and back. The more times you fly out there the more probability there is you won't come back."[32]

13

Magnificent Desolation

NASA was rushing to get to the Moon, in order to meet President Kennedy's goal. Carrying out the task on time was much more important than anything that might be found, or done, on the lunar surface. Away from Houston, however, others were dreaming of something much more sublime. In November 1967, the Reverend Terence Mangan published detailed architectural plans for a chapel on the Moon. "The moon chapel is not just one more or one other 'service' or 'gathering place,'" he explained.

> It is the reassurance of the Christian pilgrim of tomorrow, who has exchanged staff for space helmet, that there is a dimension and goal to his quest not measurable in terms of galaxies or light years alone, and the reminder to him that the greatness of his task and the glory of his accomplishment still find their fullness when brought to and offered at the Eucharistic Table of the Lord.[1]

The proposal might seem bizarre or amusing today. At the time, however, it was completely sincere. Recognizing this sincerity, it seems rather sad that a proposal so glorious should have been attached to a mission so desolate.

Christian dreamers aside, the manner in which the space race had been conducted ensured that it was never allowed to be anything other than a race. In order to attain top speed the Americans had stripped away from their space program all the cumbersome scientific and emotional baggage that might otherwise have given it purpose and meaning. Apollo was lean but empty. Armstrong admitted as much in a press conference on July 5. "The objective of this flight is precisely to take man to the Moon, make a landing there and return," he said. "The primary objective is the ability to demonstrate that man, in fact, can do this kind of job."[2]

. . .

Space was a dominating issue of the 1960s; civil rights was another. The two were distinctly separate: space showcased the country's achievement; civil rights underlined her shortcomings. The two issues did nevertheless intersect, most often when civil rights campaigners argued that the billions required to put a few men into orbit could be better utilized to help millions of blacks onto their feet. On the eve of the *Apollo 11* launch, the activist Ralph Abernathy argued that:

> a society that can resolve to conquer space . . . deserves acclaim for achievement and contempt for bizarre social values. For though it has had the capacity to meet extraordinary challenges, it has failed to use its ability to rid itself of the scourges of racism, poverty and war, all of which were brutally scarring the nation even as it mobilized for the assault on the solar system. . . . Why is it less exciting to the human spirit to enlarge man by making him brother to his fellow man? There is more distance between the races of man than between the moon and the earth. To span the vastness of human space is ultimately more glorious than any other achievement.[3]

Abernathy's complaint reached a crescendo when he led a march of perhaps three hundred followers from the Poor People's Campaign to the *Apollo 11* launch site. A light rain was falling as his army approached. A number of mules, symbols of rural poverty, were in the van, providing a stark contrast to the massive, high-tech Saturn rocket. Abernathy stopped, then gave a short speech to a crowd of onlookers who had the Moon on their mind. He pointed out that one-fifth of the nation lacked adequate food, clothing, shelter, and medical care and that, given such deep poverty, space flight seemed inhumane. The crowd remained polite, but most of the spectators wanted this spoilsport to get out of the way so that the show could start.

Abernathy was met by Tom Paine who had by his side, appropriately, Julian Scheer, NASA's public information officer. Paine's presence was carefully engineered to suggest that NASA took the plight of the poor seriously, even if it could do nothing to alleviate that suffering. He explained that he was himself a member of the NAACP and sensitive to the struggle of poor blacks. But he told Abernathy (and the assembled crowd) "if we could solve the problems of poverty by not pushing the button to launch men to the Moon tomorrow, then we would not push that button." He called the task of exploring space mere "child's play" compared to "the tremendously difficult human problems" that concerned Abernathy and other social activists.[4] As trite as that sounded, it was probably true. Then, in a clever bit of manipulation, he asked Abernathy to pray for the safety of the astronauts. Abernathy could hardly refuse. In the end, the protest had as much thrust as a Vanguard rocket.

Paine offered to admit a delegation from Abernathy's campaign to view the launch from within the area reserved for special guests. Buses were dispatched to Cocoa to pick up the group, with food packets provided. In all, one hundred men, women, and children from the campaign were admitted as honored guests of NASA. The space agency had once again worked its magic. Interviewed about the launch, Abernathy, against his better instincts, was briefly overcome by emotion. He concluded, "I'm happy that we're going to the Moon; but I'd be even happier if we had learned to live together here on Earth."[5]

"Go baby, go!" Walter Cronkite hollered as the Saturn V rose from the pad on July 16.[6] The journey to the Moon had become routine, given the success of the three previous Apollo missions. The only remaining uncertainty was the actual landing, something that could not really be practiced without actually doing it. As the lunar module approached the surface, communication was briefly lost, sending Mission Control into a spin. The difficulty was solved, but then the controller watching the fuel gauge suddenly called out "low level." Gene Kranz, the flight director in charge of the descent, recalled: "Normally by the time he calls out 'low level,' we have landed in training, and we're not even close to landing here."[7]

Armstrong, piloting the lunar module, was busily trying to find a place to park. Since he was quite keen to be able to get off the Moon eventually, he wanted to find somewhere that would give him a reasonable chance of blasting off successfully. Unfortunately, all he could see was a lot of loose rock. He started flying the thing horizontally at incredible speed. Kranz had "never seen anybody fly it this way in training."[8] The room went quiet as everyone waited. Again, the man watching the fuel gauge broke the silence: "30 seconds."

The Capcom (Capsule Communicator), Charles Duke, recalled: "All of a sudden the thing rears back and he slows it down and then comes down. And I'm sitting there, sweating out."[9] Armstrong uttered the magical words: "Contact." It was later discovered that the lunar module had between 7 and 17 seconds of fuel remaining before the crew would have had to abort.

Out on the Moon, Buzz Aldrin finished going through his checklist of post-landing chores and took out a wafer and a small flask of wine, in order to take Communion. A short passage from the book of John was read. The ceremony was entirely private, even though that was not his original intention. Aldrin had wanted to broadcast the ritual back to Earth, as the *Apollo 8* crew had done. That service had, however, evoked protests from the American Civil Liberties Union and the outspoken atheist Madalyn Murray O'Hare was suing NASA for violating the doctrine of separation of church and state.* NASA officials allowed Aldrin his private ceremony, but refused to let it go public. He nevertheless managed to sneak in his own version of a religious message when, on an open mike he said: "I'd like to take this opportunity to ask every person listening in, whoever and wherever they may be, to pause for a moment and contemplate the events of the past few hours, and to give thanks in his or her own way."[10]

Meanwhile, the unmanned Soviet *Luna 15* craft passed overhead. It had been launched three days before *Apollo 11* and had been in lunar orbit during the Americans' approach. It appears that the Soviets intended to land on the Moon, grab a soil sample, then quickly blast off, beating the Americans back to Earth. This would allow them to claim that they had landed on the Moon first and were first to bring back samples. If that was indeed the intention, it is an interesting comment on the ridiculous nature of the space race, and the obsession with making human beings the principal runners. The Soviets started that race with Laika the wonder dog, but were defeated by that same obsession. How ironic, then, that they tried, at the last resort, to cheat on rules they themselves had written. In fact, *Luna* (and irony) crashed into the Sea of Crises. Back home, the cosmonauts thanked the stars for their lucky escape.

Scott, like so many of the astronauts steeped in the politics of the space race, believes that the lunar landing was an important victory in the Cold War. "Kennedy's deadline had been met. We'd beaten the Russians. As Kennedy had said, it was the Soviet Union that had chosen space as the arena to demonstrate the superiority of its system. But we had shown them our system was better."[11] As Eisenhower had once ar-

* O'Hare had earlier instituted the action that led to the Supreme Court ruling against mandatory prayers in public schools. Her challenge to NASA did not succeed. Brian Duff, of the NASA Public Affairs Division, remarked that two things caused howls of protest from the public: praying and swearing.

gued, there were probably easier, cheaper, and more meaningful ways to make that point.

The "system" was capitalism and democracy. How precisely the lunar landing demonstrated the superiority of democracy was not clear. But as a demonstration of capitalism it was brilliant. In July 1969, it was virtually impossible to find an advertisement that did not mention the lunar landing. While NASA was careful not to be seen to be endorsing any product, it did not object when advertisers used space themes to peddle soft drinks, alcohol, cigarettes, candy, cars, airlines, panty hose, perfume and hundreds of other products. Photos of the Moon and of NASA equipment were usually provided free of charge. One of the best ads had a faithful reconstruction of the *Apollo 11* module on the lunar surface. Armstrong emerged to find the Frito Bandito had beaten him to the Moon. "Welcome to the Moon, Señor!"

Television coverage of the event was like a king-size blanket bought at K-Mart: huge, but awfully thin. CBS alone deployed 142 cameras and a crew in excess of one thousand in order to broadcast banalities to the nation. The networks competed with one another in the "expert" opinion they mobilized. Each had at least one retired astronaut. There were also poets (James Dickey and Marianne Moore), media personalities (Steve Allen, Barbara Walters, and Joe Garagiola), musicians (Duke Ellington premiered his "Moon Maiden"), scientists (Robert Jastrow and Harold Urey), and a wide assortment of other commentators (Marshall McLuhan, Rod McKuen, Arthur C. Clarke, Orson Welles, Ray Bradbury, Margaret Mead). The whole affair was like an arms race transferred to television: network producers had far more weapons than they could ever possibly use. Those who suffered were the scientists, since no one wanted unintelligible science to ruin a good story. "The producers didn't seem to have too much understanding of the science of the event," one astrophysicist complained.[12]

Writing in the *New Statesman*, James Cameron remarked ironically on how newsmen used to be criticized for sensationalizing small events. Yet here was a big event which they had "reduce[d] . . . intuitively to a sort of basic piffle . . . a level of numbing vitality." No one wanted to tackle the big questions: Why precisely did America go to the Moon, what benefits might accrue, what, realistically, might the United States do next, what did the whole thing say about America? "It could have been an elevating and eventually a self-revealing week in the history of man's lurching attempts to understand his world and himself," wrote *Newsweek* journalist Edwin Diamond. "But no one had the time or the inclination to approach meaningful material in a fresh way, to seriously consider, for example, the proposition: "We go to the Moon because we want to, we don't fix the urban mess because we don't want to."[13]

It was precisely because journalists had, for nearly a decade, faithfully followed the NASA script that the whole lunar enterprise had gone forward without serious questioning. Few critics had the guts to ask the difficult questions; those who did were dismissed as unpatriotic or Luddite by the NASA press corps. Cronkite derisively referred to the few doubters as "left-wing opposition." "I wonder what all those . . . who pooh-poohed this program are saying now," he remarked moments after the lunar landing. NASA's gratitude was expressed by Wernher von Braun, who told the press corps at the end of the mission: "I would like to thank all of you for all of the fine support you have always given the program, because without public relations and good presentations of these programs to the public, we would have been unable to do it."[14]

From July 19 to the 29th, the northeastern United States experienced virtually continuous rain. New York City had only five hours of sunshine during the entire period, flash floods hit Vermont, and the Cape Cod tourist board estimated losses at $250,000 per day. Some people blamed NASA. Quoting scripture, they argued that the bad weather was God's punishment for man's invasion of the heavens. But relief eventually came. "I've just seen the Sun on my porch for two minutes," a Bronx woman remarked when phoning a local weather station. "I think we're safe now. I thought sure, because we'd broken all the laws, that we'd never see the Sun again."[15]

In contrast, the vast majority of people were delighted by news from the Moon. In the USSR, the cosmonauts realized they had lost the race, but were still sufficiently big-hearted to admire what had been achieved. "I wanted man to succeed in making it to the Moon," Alexei Leonov confessed. "If it couldn't be me, let it be this crew, I thought, with what we in Russia call 'white envy'—envy mixed with admiration." At the time of Neil Armstrong's small step, Leonov was at the Space Transmission Corps in Moscow, where the Americans' every move was closely monitored. "Even in the military center where I stood, where military men were observing the achievements of our rival superpower, there was loud applause."[16]

Perhaps as many as one billion people around the world watched some portion of the mission on live television—the biggest audience in television history. Queen Elizabeth stayed up to watch the moon walk live. So, too, did the pope. According to the *New York Times,* the mission captured the imagination of people "from Wollongong, Australia, where a local judge brought in a television set to watch while hearing cases, to Norwegian Lapland, where shepherds tended their reindeer with transistors pressed to their ears." Trading halted on the Sydney Stock Exchange and crime rates in Milan fell by two-thirds. In Warsaw, one thousand cheering Poles packed the U.S. Embassy; in Kraków, a statue of the Apollo crew was unveiled. Across Europe, streets were deserted as crowds clustered around TV sets. The jamming of Voice of America transmissions in Cuba was briefly lifted so that the people could witness America triumphant. Everywhere, traveling Americans were stopped in the streets, congratulated, and offered free drinks.[17] Michael Fisher, an eighteen-year-old American in Moscow on a Quaker-sponsored exchange that summer, "happened to walk by the U.S. embassy that day and encountered quite a few Russians standing on the sidewalk right outside, eager to congratulate any Americans they could find."[18]

In Rome, at Christmas, a nativity scene in the Piazza Navona had a lunar module parked just behind the stable, while two astronauts, in full space gear and on bended knee, paid homage to the infant Christ. In Fife, Scotland, a boy born just after the lunar landing was named Neil Edwin Michael. Mrs. Delmar Moon of Toledo, Ohio, could not resist naming her son Neil Armstrong Moon. In Dacca, a baby Pakistani boy was named Apollo. Henry Mancini released a new arrangement of Beethoven's "Moonlight Sonata," to wide acclaim. "Moonflight" by Vik Venus reached number 38 on the Billboard chart, while "Footprints on the Moon" by the Johnny Harris Orchestra reached 39 on the Easy Listening chart.[19] Meanwhile, in Los Angeles a baker offered a line in lunar cheesecakes and a stripper in Las Vegas slowly peeled off a space suit.

The magnitude of the event unleashed a torrent of romantic nonsense. In an article in *TV Guide* prior to the launch, Cronkite remarked: "When Apollo 11 reaches the Moon, when we reach that unreachable star, we will have shown that the possibility of world peace exists . . . if we put our skill, intelligence and money to it."[20] Wernher von Braun, never one to resist hyperbole, called the lunar landing "equal in importance to that moment in evolution when aquatic life came crawling up on the land."[21] Buckminster Fuller, echoing von Braun, put *Apollo 11* at the "dead center of evolutionary events," whatever that meant.[22] Richard Nixon, eager to get in on the act, called it "the greatest week in the history since the beginning of the world, the creation. Nothing has changed the world more than this mission." That remark deeply annoyed his friend Billy Graham, who supplied, for good measure, three more important events: the birth of Christ, the crucifixion, and the first Easter.[23]

"For one priceless moment in the whole history of man, all the people on this Earth are truly one," Nixon claimed.[24] That was perhaps true, but it was just a moment. Like so many others, Nixon made the mistake of thinking that a popular spectacle was automatically an influential event. "It is the spirit of Apollo that America can now help to bring to our relations with other nations," he remarked to the American people with a crack in his voice. "The spirit of Apollo transcends geographical barriers and political differences. It can bring the people of the world together in peace."[25] At that precise moment, Jews in Palestine were doing dreadful things to Arabs, and vice versa. Vietnam was witnessing the worst violence of its long war. Contrary to all the illusions, a small step on the Moon did not erase mankind's propensity for cruelty. In any case, Nixon seems to have forgotten that without Cold War mistrust there would never have been a lunar mission.

Armstrong encouraged those who wanted to believe in an epochal event. Nearly everyone knows what he said when he first stepped on the Moon: "That's one small step for man, one giant leap for mankind." Few people know what he meant to say or, more precisely, that he left out a crucial indefinite article. He meant to say "a man," which would have given the statement some meaning, however trite. Leaving out that crucial "a" rendered the whole thing meaningless, yet those who watched the landing on TV lapped it up; Armstrong could have said "the Moon's a balloon" and millions would have praised his profundity. At one-sixth gravity, apparently, every remark sounds weighty.

Those famous eleven words were as plastic as the entire space program, something technically perfect but devoid of meaning. Armstrong provided a freeze-dried slogan, something prepared earlier and then taken into space along with the Tang and the tubes of hamburgers. The words were never meant to be deconstructed. People accepted that that one small step was a giant leap for mankind simply because Armstrong said it was so. Few paused to consider how walking around the Moon in a space suit brought progress to mankind. How, exactly did it help the starving in Africa? To what extent did it help the Vietnamese peasants who were murdering each other for the sake of an arcane struggle between communism and capitalism? Did African Americans watching the landing in a fetid shack in Alabama suddenly shout: "Golly, my life is so much better because Neil took that small step!"? Some people obviously thought so.

Armstrong's slogan was canned; Aldrin's was spontaneous. He stepped down from the bottom rung of the ladder, looked around and uttered two words: "Magnificent desolation." It's not clear whether he was trying to be profound. One suspects not; some of the best lines come naturally. It's also doubtful that he realized the wider significance of what he was saying—the fact that he was providing the perfect two-word metaphor for the space program. As an achievement, the effort to land on the Moon was magnificent. It required the energy, enthusiasm, brilliance, and creativity, not to mention bravery, of thousands. But, as a process, it was desolate, devoid of meaning. All along, NASA publicists had worked hard to give that magnificently desolate endeavor ersatz meaning. Their effort was crowned by Armstrong, who called it a giant leap for mankind. One of history's greatest ironies was that so much beauty and imagination was invested into a trip to a sterile rock of no purpose to anyone. Michael Collins, who orbited above the lunar explorers, recorded his reaction:

> The first thing that springs to mind is the vivid contrast between the Earth and the Moon. . . . I'm sure that to a geologist the Moon is a fascinating place, but this monotonous rock pile, this withered, Sun-seared peach pit out of my window offers absolutely no competition to the gem it orbits. Ah, the Earth, with its verdant valleys, its misty waterfalls. . . . I'd just like to get our job done and get out of here.[26]

As James Irwin, the *Apollo 15* astronaut, once remarked, with more incisiveness than he realized, "Astronauts come back from the Moon, say it's great, but it has no atmosphere."[27]

On the eve of *Apollo 11*'s launch, the veteran correspondent Eric Sevareid spoke of his feeling that the space adventure was nearing its denouement:

We are a people who hate failure. It's un-American. It is a fair guess that the failure of *Apollo 11* would not curtail future space programs but re-energize them. Success may well curtail them because for a long time to come future flights will seem anti-climactic, chiefly of laboratory, not popular interest, and the pressure to divert these great sums of money to inner space, terra firma and inner man will steadily grow.[28]

Norman Mailer, part of the press corps in Houston, noted how reporters didn't wait for Armstrong and Aldrin to finish their moon walk. They left the room en masse, without pausing to see what interesting rocks the astronauts might find. They had their story. The story was landing on the Moon. That's what all the effort and excitement, not to mention the sorrow and tragedy, of the previous ten years had all been about. The rest was just window dressing, an attempt to give a shallow challenge some scientific credibility. The reporters understood that the race was over.

NASA suggested as much when, shortly after the *Apollo 11* landing, they celebrated by flashing on the big screen at Mission Control Kennedy's famous words "Before this decade is out. . . ." Below was a simple message: "TASK ACCOMPLISHED, July 1969."[29] The sense of conclusion was understandable, as was the triumphalism. But the message was nevertheless deeply symbolic of NASA's greatest difficulty, namely the desire of the agency to graft its grandiose dreams of limitless space exploration onto the much more prosaic and finite ambition of the American people to win a single race.

While Armstrong, Aldrin, and Collins were traveling to the Moon, Ralph Lapp, long a critic of the space program, remarked that scientists like him were looking forward to the landing so that NASA would be able to "wind up its manned space spectaculars and get on with the job of promoting space science." Apollo, he reminded his readers, will have cost taxpayers $25 billion. "Yet manned space flights will have given scientists very little information about space. Man, himself, has been the main experiment. And man is the principal reason why Project Apollo has cost so much money." His dream, he confessed, was that upon completion of the mission, Nixon would meet the astronauts and declare: "The people of the world salute your heroism. In the name of all the people on this planet . . . I declare the senseless space race ended. And now, gentlemen, let us point our science and technology in the direction of man."[30]

Nixon was thinking along roughly similar lines, though for different reasons. Some months before, Robert Mayo, the budget director, wrote to Tom Paine asking him to provide detailed justification for a further endeavor in space, to follow the Moon mission. Paine was invited on April 4, 1969, to comment specifically on whether "the U.S. [should] undertake the development of a long duration manned orbital space station in the FY 1971–73 period." Just prior to the *Apollo 11* launch, Paine had made his ambitions clear to reporters:

> While the Moon has been the focus of our efforts, the true goal is far more than being the first to land men on the Moon, as though it were a celestial Mount Everest to be climbed. The real goal is to develop and demonstrate the capability for interplanetary travel. With some awe, we contemplate the fact that men can now walk upon extra-terrestrial shores. We are providing the most exciting possible answer to the age old question of whether life as we know it on Earth can exist on the Moon and planets. The answer is Yes![31]

Nixon's own view was that budgetary constraints, not space age fantasies, would be the determining factor in shaping NASA's post-lunar role. In his opinion,

> We must realize that space activities will be a part of our lives for the rest of time. We must think of them as part of a continuing process—one which will go on day in and day out, year in and year out—and not as a series of separate leaps, each requiring a massive concentration of energy and accomplished on a crash time table. . . . We must also realize that space expenditures must take their proper place within a rigorous system of national priorities.[32]

The battle lines were formed long before Armstrong took his first step. Those close to Nixon feared that he might be blackmailed into supporting an ambitious program because of public interest in the Moon mission. As Clay Whitehead, a White House staffer, commented, "There is significant danger that he may find himself in a very difficult situation . . . [given the] unnecessary pressure being generated by NASA in the press and on the Hill." It was important that "NASA be calmed down during the enthusiasm of *Apollo 11*, pending a systematic review this fall." Peter Flanigan, the White House link to NASA, told Paine "to stop public advocacy of early manned Mars activity because it was causing trouble in Congress and restricting Presidential options."[33]

Paine wanted $10 billion as a down payment on a mission to Mars. Mayo was thinking of a program costing one-seventh that sum, which would concentrate almost exclusively on unmanned exploratory missions to the outer planets. Mayo had not decided to veto any proposal Paine put forward, but had challenged him to justify his more ambitious program. In other words, he hoped that Paine would supply the rope with which to strangle the space program. A short time later, Paine instructed his colleagues to prepare their budget proposals for fiscal year 1971 on the assumption that they would be going to Mars as early as 1983.

Realizing that he could not depend on the congressional support that his predecessor Webb had once enjoyed, Paine desperately tried to enlist Senator Edward Kennedy, in the hope that the magic his brother had given to the space program could be revived. He asked the senator whether he would like the *Apollo 11* astronauts to take a memento from JFK to the Moon. Kennedy gave the idea short shrift. According to Paine, he made it clear that he had no interest "in identifying Jack Kennedy at all with this landing. He more or less gave me the impression that he felt that this was one of President Kennedy's aberrations."[34]

Paine did have one loyal friend in the Nixon administration, namely Vice President Spiro Agnew, who chaired the Space Task Group (STG). At the launch of *Apollo 11*, Agnew openly called for a flight to Mars. This deeply annoyed advisers close to Nixon, who felt that the vice president was painting the administration into a corner. "A Mars space shot would be very popular with many people," John Ehrlichman warned. "If the [STG] proposed it and Nixon had to say no, he would be criticized as the President who kept us from finding life on Mars." Ehrlichman, Nixon's domestic affairs adviser, had a rancorous meeting with Agnew at which he cautioned the vice president about the harm he was causing. NASA, he realized, had turned Agnew into a space zombie. "I was surprised at his obtuseness," Ehrlichman recalled. "I had been wooed by . . . the Space Administration, but not to the degree to which they had made love to Agnew. He had been their guest of honor at space launchings, tours and dinners, and it seemed to me they

had done a superb job of recruiting him to lead this fight to vastly expand their empire and budget." Ehrlichman finally had to be blunt:
"I . . . took off the kid gloves: 'Look, Mr. Vice President, we have to be
practical. There is no money for a Mars trip. The President has already
decided that.'"[35] In response, Agnew angrily demanded a personal
meeting with the president. Fifteen minutes later, after he'd been rudely
deprogrammed by Nixon, Agnew called Ehrlichman and assured him
that the Mars proposal would be removed from the STG recommendations. The final report read: "The Space Task Group is convinced that a
decision to phase out manned space flight operations, although painful,
is the only way to achieve significant reductions in NASA budgets over
the long term."[36]

Paine nevertheless kept pushing, while Agnew cheered from the
sidelines. Mayo's staff responded with a memo recommending rejec-
tion of all the ambitious proposals NASA had put forward. It concluded: "We believe the Mars goal to be much more beneficial to the
space program than to the nation as a whole." There followed a devastating critique of the whole concept of manned space travel:

> The crucial problem . . . is that no one is really prepared to stop
> manned space flight activity, and yet no defined manned project can
> compete on a cost-return basis with unmanned space flight systems. In
> addition, missions that are designed around man's unique capabilities
> appear to have little demonstrable economic or social return to atone
> for their high cost. Their principal contribution is that each manned
> flight paves the way for more manned flight. . . .
>
> NASA equates progress in manned space capability with increased
> time in space, increased size of spacecraft, and increased rate of activ-
> ity. The agency also insists upon continuity of operational flight pro-
> grams, which means we must continue producing and using current
> equipment concurrently with development of next generation sys-
> tems. Therefore, by definition, there can be no progress in manned
> space flight without significantly increased annual cost.[37]

Virtually the same argument was put forward by the Hornig and Weisner committees nearly ten years before. The memo, written just one
month after the first Moon landing, represented an astonishing out-
break of realism, a sudden emergence from a decade of somnolent fantasy. The Bureau of Budget (BoB) had mustered the courage to shout

from the rooftops that the whole manned space enterprise was self-perpetuating nonsense.

Though Paine did not give up, Nixon stood firm. A NASA budget of $3.7 billion for FY1971 was sent to Congress for approval. In response, Paine warned Nixon that his parsimony would mean that "U.S. manned flight activity [will] end in 1972 with an uncertain date for resumption many years in the future." In other words, Nixon should be prepared to go down in history as the man who had killed the dream of space travel.

Paine grossly overestimated the following NASA still commanded across the country. He failed to notice that Congress, even before the *Apollo 11* launch, was circling like a pack of hungry wolves. In response to the Agnew outburst at the *Apollo 11* launch, Senate Majority Leader Mike Mansfield had revealed that he was opposed to any Mars program "until problems here on Earth are solved." Clinton Anderson, chairman of the Senate space committee and a one-time friend of the space agency, echoed that sentiment. Margaret Chase Smith, a Republican member of that committee, felt that the government "should avoid making long-range plans during this emotional period." She believed that a Mars program would not have "the justification we had for Apollo." In the House, George Miller, chairman of the space committee, wanted to avoid binding commitments to a Mars mission and counseled a delay of five or ten years before making a decision. Joseph Karth, chairman of a space subcommittee, criticized NASA for its "complete lack of consideration for the taxpayer." Olin Teague remarked how "the easiest thing on Earth to vote against in Congress is the space program. You can vote to kill the whole space program tomorrow, and you won't get one letter."[38] Teague's observation was confirmed by opinion polls. Shortly after the Moon landing, Gallup found that 53 percent of the respondents were opposed to a Mars program, with 39 percent in favor. A short time later, *Newsweek* revealed that 56 percent of those polled wanted Nixon to spend less on space, while only 10 percent thought he should spend more.[39]

Even though Paine could not remotely understand the reasoning of legislators, he managed to sum up their attitude perfectly:

> One of the games that some people on the Hill might play would be to say, gee, let's hit the space program and wipe it out, and keep the sewers and so forth in. The idea was that, well, the reason the country was

so crummy was because we went to the moon, and by God, if we had only spent that money on all these other things that we needed to do, then we would have a great country and a crummy space program. Wouldn't it be better than a great space program and a crummy country. This was the line of reasoning they slipped into.[40]

As confirmation of Paine's complaints, Congressman Edward Koch, the future mayor of New York, confessed: "I just can't for the life of me see voting for monies to find out whether or not there is some microbe on Mars, when in fact I know there are rats in . . . Harlem apartments."[41] In the Senate, Mondale argued, in regard to the proposed Mars expedition, that "it would be unconscionable to embark on a project of such staggering cost when many of our citizens are malnourished, when our rivers and lakes are polluted, and when our cities and rural areas are dying. What are our values? What do we think is more important?"[42] Senator William Fulbright, citing the rough time the administration's NASA budget had in Congress, explained, very simply: "We voted for sewers. Certainly sewers are more important than going to the Moon."[43]

Congress eventually settled on a budget of $3.269 billion for 1971. The budget would hover around the $3 billion level until the mid-1980s, in real terms one-third of its peak in the 1960s. The figure was a political compromise between the pragmatic concerns of the BoB and the grandiose dreams of NASA. Nixon would be able to keep Paine's ambitions in check without at the same time being branded as the man who killed the space program. Billions would continue to be spent in homage to a dream, but that dream would never be realized.

Paine held on to his fantasies, all the while hoping that the country might renew its romance with space. In June 1970, shortly after the release of the 1971 budget, he called a three-day conference on the future of the space program. Among those attending were the stars of the golden years, including Wernher von Braun, Arthur C. Clarke, Robert Gilruth, and Neil Armstrong. Paine made it clear that participants were not to keep their fantasies in check, he wanted them to extend their imagination in every conceivable direction, in an attempt to draw blueprints for what he called "Spaceship NASA." New types of engines would be built that would not only extend the range of space travel but also enable the construction of an "Intercontinental Space Plane" able to reach any point on Earth in less than an hour. Advances in food manufacture would allow the synthesization of food from fossil fuels, thus "freeing man from his 5000 year dependence upon agriculture."

Paine wanted the group to see NASA like a grand fleet, exploring mysterious, dangerous, and uncharted waters, for adventure, discovery, and profit. "Consider NASA as Nelson's 'Band-of-Brothers'—Sea Rovers—combining the best of naval discipline in some areas with freedom of action of bold buccaneers in others—men who are determined to do their individual and collective best to moving the planet into a better 21st Century." He was delighted with the conference. Afterwards, he wrote enthusiastically to Nixon: "The results are exciting and I would like to request an appointment to present to you our best current thinking. . . . The purpose is . . . to give you a heretofore unavailable Presidential level long-range view of man's future potential in space."[44] Nixon, not about to expose himself to the space bug, politely thanked Paine for his submission.

A short time later Paine went back to work for General Electric, where he had been employed before joining NASA. He explained that he needed to think practically about the problems of raising four children on a government salary of $42,500 per year. At some level at least, he knew the value of a dollar.

The race was won. But, in order to give the impression that it was always more than just a race, Americans would go back to the Moon. NASA would invest the missions with further scientific hocus-pocus, again to give the impression that it was a very serious pursuit and to convince the public that something worthwhile (perhaps even profitable) might come out of the great adventure—something more than magnificent desolation.

John Kennedy had said that Americans should land on the Moon before the end of the decade. They'd done that. But wouldn't it be neat if they were able to complete the task *twice* before the new decade began? This was, after all, the Cold War, where propaganda victories took the place of military advance. To go twice would put an exclamation point at the end of a bold statement about American capitalist supremacy.

The *Apollo 12* mission was therefore scheduled for November 14, 1969. Unfortunately that's winter, something that happens even in Florida. The mission to the Ocean of Storms began in storms at Cape Kennedy. According to the NASA rulebook, rain was okay, but lightning was not. Since the meteorologists saw no chance of lightning, Mission Control, aware of how narrow were these windows of opportunity for going to the Moon, decided to press ahead. The Saturn V rocket roared to life, but before it could clear the launch tower, lightning struck. Circuit breakers blew, and, inside the command module, the astronauts were momentarily plunged into darkness. "We just had everything in the world drop out," Pete Conrad told a momentarily bewildered Mission Control.[45]

A less complicated vehicle would certainly have crashed. But the entire package had concentric layers of redundancy built into every system. Almost immediately, backup power kicked in, and the rocket stayed on course. "It's a matter of making every possible human effort to avoid failure of a part," von Braun once boasted. "And then taking steps to avoid the effects of a failure if one should develop anyway."[46]

After a thorough check of the systems, Pete Conrad, Richard Gordon, and Alan Bean continued on their way to the Moon. Conrad and Bean landed in the Ocean of Storms and found, to no one's surprise, a desolate monochromatic landscape full of rocks. As a result of some nifty navigating, they landed very close to the Surveyor 3 lunar probe, which had touched down a few years earlier. They retrieved the camera and a few other pieces of the equipment, so that engineers back on Earth could study the effects of the lunar environment on man-made objects. To everyone's surprise, a microscopic biological organism was found on the equipment when it was studied back on Earth. There followed a brief flurry of speculation that perhaps life had been found on the Moon. That possibility was quickly swept aside; in the end, the simple explanation seemed most logical: Bean and Conrad had probably contaminated the materials by touching them with dirty fingers. In other words, there was no life on the Moon, just rocks.

The popularity of the rock-collecting expeditions was quickly waning. *Apollo 13* was scheduled for April 1970. The mission was given to James Lovell, Ken Mattingly, and Fred Haise. Just days before the launch, fellow astronaut Charles Duke was exposed to measles. He in turn unwittingly exposed the flight crew. Haise and Lovell had immunity, but Mattingly did not. By this stage, NASA should perhaps have contemplated the unlucky reputation of number 13, but it does not do for an agency specializing in science to be superstitious. The show had to go on. Mattingly was replaced by Jack Swigert.

The rest of the story is well known, thanks in large part to the film starring Tom Hanks. An explosion ripped through the outer skin of the Command Module, which quickly lost electrical power. To add to their woes, Haise came down with a serious infection and was desperately ill for most of the journey. The crew had to use the lunar module as a lifeboat, up to reentry. They survived through a combination of courage, flexibility, determination, and imagination, not to mention sheer brilliance—both in the spacecraft and on the ground.

A subsequent investigation revealed that the cause of the problem dated back to 1965, when the power supply on the Apollo spacecraft was increased from 28 to 65 volts. The manufacturer of the suspect oxygen tanks had failed to make an adjustment for this change, which means that every previous Apollo mission had flown with the same potentially disastrous malfunction. In this case, however, the problem was compounded because the tank in question had been damaged in preflight exercises when it had previously been installed on *Apollo 10*. It was replaced, repaired, and put on *Apollo 13* by technicians who clearly were not superstitious. The tank continued to cause problems in the preparation for the *Apollo 13* flight, but no one considered them serious enough to cancel the flight. So much for von Braun's assurances about removing all conceivable faults.

The film *Apollo 13* is a masterpiece, a riveting drama of human endurance under the most challenging circumstances. But the drama obscures some harsh realities about the space program. The crew undoubtedly exhibited immense courage and were helped by brilliantly creative improvisation on the ground. Survival was first and foremost a triumph of human ingenuity. But while humans had solved the problem, they were also the cause of it. A crisis developed because men were on board, which made it imperative that they be brought back. An unmanned flight would simply have been aborted and the only loss would have been millions of dollars worth of equipment.

Another reality is even more telling. Of all the lunar missions, probably 99 percent of Americans can recall only two: *Apollo 11* and *Apollo 13*—the first one and the nearly disastrous third one. The others have faded into obscurity and insignificance. At the time, they seemed a bit like a scratched record endlessly repeating itself. With each launch, attention faded, telecasts shortened, and criticism of the costs grew more strident. Before disaster struck, one of the transmissions from *Apollo 13*, which was supposed to be aired live on television, was can-

celed by CBS in favor of the *Doris Day Show*. After the explosion, however, the mission became a media event because it was a failure. Success was no longer news. Flight director Gene Kranz later became famous for his immortal line "failure is not an option." In truth, however,
the entire mission was a failure, since its purpose was to put two more
men on the Moon. The fact that it failed made it interesting and newsworthy.

If *Apollo 14* is remembered at all, it is for the fact that Alan Shepard
hit some golf balls on the lunar surface. In a subsequent *All in the Family* episode, Michael "Meathead" Stivic (Rob Reiner) argued with his fa-
ther-in-law, Archie Bunker (Carroll O'Connor), about the cost of the
space program. "You don't think," he shouted, "we got anything more
important to do with twenty billion dollars than to send a guy up to the
Moon to hit a few golf balls?"[47] A majority of the nation agreed. During
the *Apollo 15* mission, an anonymous viewer phoned his local TV station to suggest that a large rock discovered by the astronauts should be
named in honor of "a taxpayer selected at random from the computers
of the Internal Revenue Service."[48]

David Scott, who traveled on *Apollo 15*, believed wholeheartedly in
the value of the space program.* To him, it was not just symbolic of the
superiority of the American system, it was also a sublime demonstration of mankind's need to explore. "As I stand out here in the wonders
of the unknown at Hadley," he said while on the surface of the Moon,
"I sort of realize there's a fundamental truth to our nature. Man must
explore. And this is exploration at its greatest." Perhaps for some men
that was true, and certainly sincerely felt. But Scott's description of the
Moon does make one wonder why there was ever a need to go there at
all, and, even more so, to return:

> To an "Earthling" one of the Moon's most striking features was its still-
> ness. With no atmosphere and no wind, the only movements we could
> detect on the lunar surface, apart from our own, were the gradually
> shifting shadows cast to the side of rocks and the rims of craters by the
> Sun slowly rising higher in the sky. There were no other features: no
> trees, bushes, rivers, streams, flowers, grass, animals or birds—none of
> the signs of nature that human beings have evolved with and are used
> to. There was no sound, either, apart from the gentle humming of the

---

* His was the first mission to deploy the hugely expensive Lunar Rover.

equipment in our backpacks. There were no clouds, haze or mist, and there appeared to be no color. The sky was pitch black except for the deep blue and white of our own planet suspended high in the sky like a Christmas tree ornament.[49]

Astronauts always comment upon how lovely the Earth is from space, especially in comparison to the barren Moon. The view is fantastic. From that altitude, Earth is a lovely pearl dappled in shades of blue, pink, and white. It's pretty precisely because one can't see the hunger, the cruelty, the pollution, and the hatred that despoil the planet. Illness, poverty, crowded schools, and homeless people—all the problems crying out for money—magically disappear.

Since the missions, memory has faded almost completely. Some people can recall a very expensive lunar rover, but only the space geeks have a real sense of the distinct nature of each mission. That is entirely understandable given that NASA and the public were dancing to a different tune. There was hardly a whimper of protest from viewers when the three main networks kept coverage of *Apollo 17,* the last lunar mission, launched in late 1972, to a bare minimum. PBS briefly considered stepping into the void but in the end could not justify the huge expense for an audience so small. When CBS cut the last seventeen minutes of *Medical Center* in order to show the launch, the network was bombarded with complaints from viewers. ABC gave only half an hour to the lunar landing, but did include a short report during the halftime of the Jets-Raiders NFL game.[50] Instead of showing the final steps of man on the Moon, NBC decided to broadcast a repeat of the *Tonight Show* with Johnny Carson. "We . . . were already yesterday's news," Cernan remarked bitterly.[51]

For most Americans, the race was over when Aldrin, Armstrong, and Collins returned. There seemed little sense in going back. NASA, on the other hand, was still thinking in terms of the Moon missions as a prelude to something bigger. But by this stage the agency had neither the imagination, nor the money, nor the mandate to figure out what to do next.

So what was the lunar mission all about? What was it for? Charles Schultze, LBJ's budget director, had to deal with the financial implications of Kennedy's challenge. In the end, he decided that there were worse ways to spend money:

Looked at as entertainment, it sure did have its entertainment value. . . . It's not gonna give us much scientific knowledge, you get some rocks back from the Moon. But you do it on a per capita basis, and as an entertainment tax it's a great entertainment. You get not only the Moon landing itself, but you got all those initial, you know—the first shot around the Earth, Glenn, all of that business. And then you get—I don't remember how many—four or five moon landings and, eh, as an entertainment tax, per capita, it wasn't bad.[52]

Walter Cronkite always seemed to have the final word on so much of what happened in, or to, America. He watched the space race at close quarters from Sputnik to the last journey of the lunar rover. Was it worth it? You bet.

The success of our space program . . . in that terrible decade of the 60's, played an important part in maintaining a semblance of morale in a country that was very, very depressed in everything else that was happening. . . . The Civil Rights fight, the assassinations, the Vietnam War, these were things that split America in a way that we hadn't been split since the Civil War of the 1860's. And here was this one program where people could look up and dream if you please of incredible adventure. And there was a pride in that. It had a great deal to do with maintaining some sense of balance in this civilization of ours.[53]

In other words, the lunar mission was a $35 billion happy pill administered to a generation of depressed Americans. There's no doubt that the 1960s were tough years, but as a cure for depression, surely Valium would have been cheaper. Cronkite was a good man and a great American. But every once in a while, he did talk nonsense. In 1974, he argued, in defense of new space goals: "How much is it worth to prove in an era of cynicism and gloom that man can do anything he wants to do as long as he has the will to do it and the money to spend."[54] At one time, that sort of remark sounded inspiring. In the lean 1970s, it merely sounded arrogant.

Von Braun stopped smiling when the world stopped listening. "It's essentially like telling a long range planner that his plans were shit," his protégé Jesco von Puttkamer exclaimed.[55] Von Braun quit NASA and went to work for Fairchild, selling helicopters to countries in South America. There were a lot of Germans down there who still revered him. He died of cancer in 1977 at the age of 65, a broken man who had once taken America on a ride to the Moon.

14

Nothing Left to Do

Going to the Moon was supposed to be the first step. Where Apollo went, ordinary people would follow. In 1968, PanAm began to take bookings for its *Clipper* flights to the Moon, scheduled to depart in the year 2000. One of the first to reserve a seat was Governor Ronald Reagan. Executives at Hilton Hotels meanwhile began musing about the possibility of building an underground resort on the Moon. A confident Tom Paine predicted that "by 1984 a round trip, economy-class rocket-plane flight to a comfortably orbiting space station can be brought down to a cost of several thousand dollars," while a trip to the Moon would run in "the $10,000 range."[1] Herman Kahn took time out from musing about nuclear Armageddon to predict that the lunar surface would become a popular honeymoon destination by 2029. There'd be no reason to pack luggage since everyone would be wearing disposable clothes.[2]

Nearly eight years after the last lunar mission, Reagan was elected President of the United States. One of his first major acts was to deregulate the American airline industry. This placed huge pressure on the big carriers. An early casualty was PanAm, which never got to launch its *Clipper* service to the Moon.

Twelve men walked on the Moon. Privately they refer to themselves as the Order of the Ancient Astronauts. Today, the quest to go to the Moon seems as strange as stuffing fraternity brothers into phone booths, swallowing goldfish, or listening to the 1910 Fruitgum Company. While both Presidents Bush have tried to resuscitate enthusiasm for space with bold challenges to go to Mars, the American public's reaction has been decidedly lukewarm.

The lunar mission was a very cold thing, cold as the dark side of the Moon. There's not a lot of emotion in the press conferences Armstrong gave before or after his mission. Good old Ham would have screamed and flared his teeth, while Enos might have masturbated. Armstrong, in

255

contrast, spoke in gray monotones about technological parameters and energy coefficients. When asked if he would keep a piece of the Moon as a memento, he replied: "No, that's not a prerogative we have available to us." "What a joy it could have been," Norman Mailer remarked, "to cover this Moon landing with a man who gave neat quotes, instead of having to contend with Armstrong, who surrendered words about as happily as a hound allowed meat to be pulled out of his teeth."[3]

Whatever warmth and emotion existed within the mission was found back on Earth. It was usually manufactured by NASA PR agents and disseminated by dutiful journalists eager to play their part in the biggest media event in history. But that emotional high could not be sustained, since for the public the mission was a self-contained quest worthy of neither repetition nor extension. Even the most dramatic film loses impact on second viewing. NASA could never maintain the drama that its PR agents (and the cooperative media) provided. Eventually, reality intruded, and the space quest was revealed as a rather boring enterprise of little real value to ordinary people on Earth.

Adventures are not supposed to be endless. As Sevareid predicted, NASA was eventually defeated by its own success. It was judged for what it was: an effective way to get to the Moon quickly. It could not supply justification beyond momentary practicality; it could not sustain the public's brief romantic attachment to space travel. After the moon-dust from *Apollo 11* settled, economic considerations became predominant. Lunney understood precisely why the space age came to an end. NASA, he feels,

> had accomplished its original purpose, and the return, although seen as very valuable by the lunar scientists, was not seen as so valuable to be worth that kind of money by the political system. There wasn't a lot of purpose to be served by just continuing to do the same thing over and over again. . . . You know, other endeavors were different. For example, the opening up of the New World. There clearly was, once people got over here, a reason to come back.[4]

Those who justified the presence of men in space argued that the early astronauts were like the medieval seafarers, looking for places to colonize. But the efforts of Columbus and Magellan were inspired by the commercial potential of new territories—exploration was pointless unless commerce followed. The Portuguese and Spanish courts would have pulled the plug on the explorers quicker than you can say Vasco da Gama if their voyages had been exclusively esoteric, or if they had brought back only worthless rocks. Instead, they returned with valuable commodities—precious metals, spices, trinkets, potatoes—which thrilled the medieval money crunchers. In addition, the places they sought to explore were, by virtue of their existence on Earth, actually habitable. The same could not be said for colonies on the Moon or Mars. Even if the gargantuan problems of mining the Moon for precious resources could eventually be solved, it seems only logical that any such effort would, by necessity, be a strictly utilitarian affair, not a long-term settlement with saloons, banks, and brothels. The Moon, remember, makes Antarctica seem like an oasis.

The lunar mission was a historical accident. It didn't make sense financially, as previous voyages of exploration had. But for a brief moment in the twentieth century, money didn't matter. The Cold War was not a good time for accountants; the important thing was to score points against the Russians, and to do so was priceless. At the very moment when financial scruples were temporarily abandoned, it also happened that the technology was available to provide a way to get to the Moon. But now, even though technology has progressed, the political reasons for such expensive ventures have disappeared and justification is difficult to find.

Buckminster Fuller once claimed that space exploration was as natural as "a child running around on its own legs." Neil Armstrong, barking up the same tree, explained it in terms of genetic inheritance. The astronauts, he said, were driven to outer space "just as salmon swim upstream."[5] But those who explained the mission in terms of "man's need to explore" have been unable to explain why the decline of the space program has not, as logic would suggest, led to cultural bankruptcy. Bear in mind that salmon who fail to make it upstream do not reproduce. They die without replicating themselves.[6]

The "need to explore" is in fact a great myth, an imagined construct used to fleece the taxpayer who gets vicarious adventures instead of hospital beds. The quest to explore might inspire a few people who feel the need to climb Everest or walk to the North Pole, but it is not an indicator of cultural vitality. The malady is most common among those who lack the imagination to enjoy life on Earth or to see how it might be improved. The quest is not by itself important enough to explain why an entire society should spend billions in order to collect a few red rocks from Mars.

Professor Amitai Etzioni, who howled about the space program's waste of money back in the 1960s, took delight in the fact that Americans finally came to their senses. But he still battles against NASA, its talent for spending money, and the grip it exerts. "Although rarely discussed . . . the greatest achievements of the space program—whether by NASA, the military, or the private sector—have been the result of unmanned vehicles and instruments," he argues. "And, by the way, most of those were in near, not deep, space. Thus we have achieved giant improvements in worldwide communication, navigation, mapping, weather forecasting, and above all surveillance by unmanned satellites." In contrast, manned flights into deep space have brought back little of enduring value. The Moon did not become the military high ground that Johnson predicted. It is not a treasure trove of natural resources. Scientific experiments conducted on the lunar surface have not improved the quality of life on Earth. Nor has the exploration of space done anything for the enhancement of our collective soul.

The great tragedy of the effort was that the best of American technology and billions of American dollars were devoted to a project of minuscule benefit to anyone. Armstrong's small step did nothing for mankind. One year after his return to Earth, even he had begun to lose faith in the idea of a giant leap. When asked whether he still believed that the conquest of space would render war obsolete, he replied: "I certainly had hoped that point of view would be correct . . . [but] I haven't seen a great deal of interest, or evidence of that being the truth in the past year."[7] One can speculate endlessly on what might have happened had all that effort and money been devoted to an earthly project of more obvious benefit to the world, such as medical research or the development of alternative energy sources. But such speculation is pointless, since in the mood of the 1960s a purely humanitarian project offered no propaganda value. In the Cold War, no points were awarded for doing good. The contest had by this stage become so abstract that outer space provided the only worthy arena for conflict.

James Webb, in his *Space Age Management*, argued that the Moon mission proved that Americans had the "ability . . . to accomplish almost any task" they might set for themselves. He concluded that there was "no reason why we cannot do in other areas what we have done in aeronautics and space," and went on to argue that in fact the NASA system should perhaps "become the pattern needed by the nation."[8] All that sounded very nice, but in truth it was hogwash, a lame attempt at self-justification. There were quite a few reasons why Americans could not accomplish on land what they had accomplished in space. The Moon mission was not a demonstration of American ability, but rather of the nation's tolerance—it was a measure of what Americans would allow to happen with their efforts and their dollars. They achieved a magnificent success in space because they managed for a brief period to agree that space exploration was important, even necessary. In contrast, the really crucial things that needed doing back on Earth were simply too contentious to inspire that kind of agreement. Only in outer space could consensus be reached.

When the money dried up, and the space program turned inward, the hard-core fantasists, those who had always seen the lunar mission as simply the first step on the road toward extraterrestrial colonization and limitless exploration, took over. Their dreams were no longer held in check by NASA's necessitous practicality. They could fantasize at will, never having to worry about what the laws of physics might deem possible or economists might consider affordable. Space now belongs to them. They explore it in their dreams, which is the only place where travel at speeds faster than light are possible. They talk to each other in geek conferences and on Internet chat rooms, always cursing the small-minded public who pulled the plug on Apollo.

When the space age ended, the alien age began. In the early 1990s, the Disney Corporation decided to close down its Mission to Mars ride, itself a direct descendant of the Rocket to the Moon attraction Wernher von Braun had helped to design. In its place came Alien Encounter, in which an extraterrestrial stows away on a spaceship. This made things easier for Disney, as one executive admitted: "One way for an attraction to remain timeless is for it to be based in fantasy, rather than reality."[9]

Star Trek and *Star Wars* boldly go where NASA, stuck in reality, can never travel. Laws of nature do not apply in the imagination. In Hollywood, spaceships are blindingly fast; their range is determined only by the captain's ambition. Planets are beautifully threatening environments supporting all manner of weird aliens, not baked and barren rocks. Aliens are everywhere—real, anthropomorphic beings—not microscopic organisms in a primal soup or fossils in lumps of ancient ice.

Mr. Spock, Captain Kirk, and Obi-Wan Kenobi don't have to wear clumsy space suits or eat goop from tubes. They demonstrate that man can conquer space, instead of space conquering man.

The line between fiction and fact is thinly drawn. Fans of Spock usually also believe in the possibility of extraterrestrial life. According to a poll conducted in 1985, 43 percent of Americans believe that UFOs, carrying aliens, have visited the Earth (the same poll found that only 47 percent accepted the theory of evolution).[10] Prior to Sputnik, few people outside Roswell ever gave a moment's thought to UFOs. Reported sightings rose sixfold in the three months after Sputnik and have remained high ever since. The sudden increase in sightings after 1957 is proof of the power of suggestion. These aliens, who come from very far away, seem to have figured out the problem of traveling at speeds faster than light. They are, in other words, superior beings, much further on the evolutionary scale than we are. But most of them still look like us, except for the occasional frightening addition to their physiognomy. The ones in *Star Trek* have heads and arms, chests and legs, but usually some distinguishing feature like overdeveloped chins or massive cranial protrusions—as if they got caught in a plasticine storm.

By far the most popular alien type is that known in ufologist circles as the "Gray." He's got doelike eyes, an aquiline nose, and a large wobbly head with pointy chin, stands about five feet tall, and has no apparent musculature and no clear sexual orientation. Gray is reassuring because he's not physically threatening—in a straight fight, almost any human being could beat him senseless. But what he lacks in strength, he makes up in intelligence.

He's so intelligent that, despite his millions of visits to Earth, he's never once left behind a shred of hard physical evidence of his presence. Not a single footprint, gum wrapper, interstellar wrench, or mound of green alien excrement. The point is that outer space, as we imagine it, doesn't exist. It's a product of this Earth, a figment of human imagination. The truth isn't out there. It's right here.

"The legacy of Apollo has spoiled the people at NASA," a disappointed Wernher von Braun once remarked. "They believe that we are entitled to this kind of a thing forever, which I gravely doubt. I believe that there may be too many people in NASA who at the moment are waiting for a miracle, just waiting for another man on a white horse to come and offer us another planet, like President Kennedy."[11] Isaac Asimov made a similar point:

> I hope the Russians would do something like Sputnik to us again and wake us up. After *Apollo 17*, it became clear that the attitude toward expeditions was "We scored a touchdown, we won the game, now we can go home." It struck me as stupid thinking. Then again, I began to think, that's how it was presented to the people all along. There was hardly any concentration on the gains in technology or the joy of discovery. It was a matter of beating the Russians. On the other hand, if it was not presented that way, one wonders if Congress would have provided a red cent.[12]

Webb, who guided NASA through the glory years, never really liked the idea of a race to the Moon. A race, he understood, was finite—once you cross the finish line, you stop running. One of his most important tasks at NASA, as he saw it, was to encourage his colleagues to eschew limits. "It is an interesting fact," he once remarked, "that the lunar landing represented a requirement that we learn to use 98 per cent of the energy it takes to keep going to Mars or Venus."[13] Buried within that remark one finds typical NASA logic: the lunar missions had intrinsic meaning, but their meaning and value will increase exponentially if they're used as stepping-stones to further explorations. And those further explorations—well, you get the picture. What results is an endlessly repeating formula for spending billions of dollars, just as Jerome Weisner warned back in 1961.

But where to go after the Moon? In 1985, President Reagan appointed Tom Paine to head a National Commission on Space with a remit to prepare a report on the future of space exploration. Paine used the opportunity to remarket the fantasies he tried to sell to Nixon back in 1969. The lavishly illustrated report, entitled *Declaration for Space,* was released in May 1986. It called for a "pioneering mission for 21st century America," in other words, another Kennedy-style mandate. The United States should "lead the exploration and development of the space frontier, advancing science, technology, and enterprise, and building institutions and systems that make accessible vast new resources and support human settlements beyond Earth orbit, from the highlands of the Moon to the plains of Mars." The plans were bold and enormously costly, but the cost wouldn't matter—"in view of America's projected economic growth," it was a virtual certainty that "the percentage of the US GNP invested in opening the space frontier would remain below one-half of the percentage spent on space during the peak Apollo years." In any case, most of the cost would be borne by the private sector, since industry would be keen to reap the enormous profits to be made. Eventually, the report confidently predicted, "Space travel will be as safe and inexpensive for our grandchildren as jet travel is for us."

What the authors failed to understand was that the space quest had appeal when it was an adventure—when each mission held the possibility of disaster and when astronauts seemed clearly heroic. Making spaceflight routine removes its main attraction, as the sorry history of the Space Shuttle has demonstrated. Ignoring all this, the group plowed on with its fantasies:

> The immediate benefits from advances in science and technology and from new economic enterprises in space are sufficient in our view to justify the civilian space agenda we propose. However, we believe that the longer-term benefits from the settling of new worlds and the economic development of the inner Solar System will prove even more rewarding to humanity. These returns are difficult to quantify. What was the true value of developing and settling North and South America, Australia, and New Zealand? Today more people speak English, Spanish, and Portuguese in the New World than in Europe, and they have built economies surpassing those of Europe. But the contributions to humanity from Columbus' "New World" are surely far beyond its material returns, impressive as they are. We believe that in removing terrestrial limits to human aspirations, the execution of our proposed space agenda for 21st-century America will prove of incalculable value to planet Earth and to the future of our species.[14]

A lot of intelligent people worked on the *Declaration for Space*, including the diplomat Jeanne Kirkpatrick, the Nobel physicist Luis Alvarez, Neil Armstrong, Chuck Yeager, Gerard O'Neill, and Paine, yet it reads like it was written by Captain Kirk. Reagan praised it, then filed it.

The idea that a better world exists and can be discovered in outer space continues to fuel fantasy. Echoing Konstantin Tsiolkovsky, Paine envisaged a brave new world that would follow on the heels of Armstrong's small step. "As with the American experience of 1776, founding a new society in a demanding environment will sweep aside old world dogmas, prejudices, outworn traditions, and oppressive ideologies. A modern frontier brotherhood will develop as the new society works together to tame its underdeveloped planet for posterity." All that was needed was for America to summon "the vision and determination to press forward in *Apollo XI*'s fiery wake."[15]

In 1981, O'Neill, self-appointed guru of space, published his prophecy entitled *2018: A Hopeful View of the Human Future.* In it he forecast that spaceflight would usher in "perpetual plenty," in the process solving the population crisis, the energy problem and, for good measure, international conflict.[16] Poverty would be completely abolished. O'Neill's vision of an outer space Utopia consisted of what could only be described as a cosmic cruise liner, a gigantic spinning top in which a few hundred carefully chosen individuals would create the perfect community—rather like the Brick Moon. His Island One, the first stage of this massive vision, would, he forecast in 1976, cost $31 billion to make and could be up and running before the end of the twentieth century. "With an abundance of food and clean electrical energy, controlled climates and temperate weather," he promised, "living conditions . . . should be much more pleasant than in most places on earth."[17] Life up there would be Earth-like, but much better, because, as we all know, people will behave in outer space. Life would be reassuringly simple, predictable, and safe, with more time for the enjoyable and enriching things. The intrepid orbiting colonists would enjoy the kind of amenities "which we would expect in a small, wealthy resort community on Earth: good restaurants, cinemas, libraries, perhaps small discotheques."[18]

O'Neill's fantasy was cleverly rooted in science and hard-nosed accounting. For this reason, it could not be easily dismissed. He supplied figures to suggest that, thirteen years after the establishment of Island One, it would be supplying the entire energy needs of the United States, by collecting solar power and beaming it back to Earth. This got NASA interested. Their calculations caused the price tag to rise to $140 billion, but this still didn't dampen interest, given the huge energy savings promised. The agency gave O'Neill $40,000 a year for five years to keep thinking.

The other thing that got the blood racing was the fact that Island One, that most artificial of environments, seemed to chime so well with environmentalists, not just because of the clean energy it would provide, but also because what O'Neill proposed was a seventies-style, self-sufficient, pass the granola, eco-commune. By this stage, he was so excited by his perfect society in space that he predicted in *The High Frontier* that by 2050 there'd be more people living in space than on Earth. This would allow the Earth to return to its wild, pastoral virginity. A favorite rallying cry of O'Neill's fans went: "Declare the Earth a wilderness area: if you love it, leave it."[19]

O'Neill nevertheless had his critics. Lewis Mumford called his ideas "technological disguises for infantile fantasies." "I see acres of air-conditioned bus interior, glinting, slightly greasy railings, old rivet heads needing paint," the inventor Steve Baer complained. "I don't hear the surf at Carmel and smell the ocean—I hear piped music and smell chewing gum. I anticipate a continuous, vague, low-key 'airplane fear.'"[20] Recently, an anonymous critic posted the most damning criticism of O'Neill on an Internet forum:

> The romance of pioneering in space will wear off real quick once the public realizes there are no holodecks, or green bountiful parks in Moon domes, or exciting battles with fierce alien invaders, just 10 cubic meters of cramped quarters, recycling your own urine, living in your own sweat, facing the dangers of decompression daily, seeing the same goddamn faces every day, knowing that down below on Earth people are laughing, and loving, breathing fresh air and running in spring sunshine, meeting people and living life.

The correspondent went on to argue that, if space were ever to be colonized, the only people willing to go would be those who today "risk their lives on shaky home-made rafts on the open sea in search of a better life." Those with nothing to lose might just be able to put up with life in a spinning aluminum can. "They have far more in common with the passengers on the *Mayflower* than the overfed, overweight, self-satisfied, self-righteous, tax-paying, groundhogs sitting behind their computers and filling the Net with wannabe hogwash, boasting about their ancestors' deeds and imagining themselves to be their spiritual heirs."[21]

In the end, it wasn't the hollowness of O'Neill's fantasies that killed his project—the fact that his imagined world would be unimaginably horrible—but rather that technology could not support his fantasies. He had extrapolated from what had been achieved in ten short years from 1959 to 1969, and assumed that technology would continue on an exponential curve, making any vision capable of realization. But the limitations of technology were revealed by the Space Shuttle, which did not even live up to the narrow visions of its designers. They promised something quick, reliable, cheap, and reusable but produced something slow, expensive, dangerous, and extremely complicated. Since something like a shuttle was the first small step in O'Neill's gigantic vision, reality hit hard. O'Neill never understood how the lunar quest was fueled by Cold War politics. Remove the antagonism and technology goes elsewhere.

Defenders of the space program often argue that man will someday need to find an alternative home in which to live, after the resources of the Earth have been exhausted. On the twenty-fifth anniversary of the lunar landing, George Mueller argued:

> We have today a clear reason for spreading humanity to other planets. The sight of the awesome explosion of a relatively small asteroid on the surface of Jupiter vividly illustrates the vulnerability of our planet to such a catastrophe. It is time for us to protect . . . our life forms from a similar catastrophe on Earth, a catastrophe which we know once destroyed much of life on this planet.[22]

The Earth is indeed doomed, but where precisely might refugees eventually go? The solar system certainly offers no habitable alternative. The idea of migration is really only taken seriously by demented dreamers and sci-fi buffs who conveniently ignore the fact that the universe is rather large. Travel, even at immense speed by a select few, would take impossibly long. Suppose a spaceship could be built capable of a speed of one million miles per hour. (The Apollo capsule reached a maximum speed of about 50,000 miles/hour.) If the destination were the nearest star system theoretically capable of harboring planets, that super speedy spacecraft would require 4,000 years to reach its destination.

The problem is solved in science fiction by spacecraft going faster than the speed of light. The *Enterprise* is one very fast ship. But that kind of speed is only possible in the human imagination. Assuming Einstein was right, an increase in velocity implies an increase in mass. The effect is negligible at the relatively slow speeds that a Saturn rocket traveled. But at velocities approaching the speed of light, that effect would be prohibitive. The heavier the craft, the more energy necessary to propel it. At the speed of light, the craft's mass would be so great that its energy requirement would be infinite.

In other words, even if a perfect world exists in deep space, how do we get there? Furthermore, how do enough of us get there for the trip to qualify as a giant leap for mankind? What the space age has demonstrated is just how hard it is to get anywhere in space. Rather than expanding man's horizons, it exposed his limits. In galactic terms, the Moon is just a spitting distance away. Only the dreamers and fantasists are able to solve the one all-consuming problem of the universe—its size.

The Moon was part of the sixties zeitgeist. For a very brief period it was in vogue, like lava lamps, love beads, and bell-bottom trousers. In the 1960s the race was symbolic of technological advance. Since technology was assumed to be benevolent, so too was it presumed that the fruits of technology (for instance, the Moon itself) would benefit mankind. Reaching the Moon would be profitable, either for the bank balance or for the soul. The quest turned into a contagion. America, certainly, and the world, to an extent, was briefly moonstruck. But exposure to an illness often leads to the development of an immunity. We've recovered from the disease, and most of us at least are no longer susceptible.[23]

Space enthusiasts always thought that exploration was the destiny of man. They thought it was the future. But they were wrong. The Moon was the present and is now the past. It was simply a battlefield in the Cold War. Since that war remained cold, without real battlefields of bloody conflict, symbols and propaganda became enormously important. At some point it was decided that the ability to put a man in orbit around the Earth (or perhaps send him to the Moon), and bring him back again, was a legitimate symbol of a nation's virility. With the end of the Cold War, the space race lost its raison d'être. The launch pad was pulled from under the rocket. Without the political justification, the shallow nature of space exploration (and, in particular, man in space) was revealed. Its limitations were no longer hidden behind the egotistical quest for prestige.

Landing on the Moon did not change the world. It did not bring a more perfect, harmonious community of nations. It was not a giant leap for mankind. In answer to those who predicted that the lunar landing would usher in an era of peace, love, and international brotherhood,

Mumford reminded Americans that the space program was nothing more than "an extravagant feat of technological exhibitionism." We need to be reminded, he wrote, that

> any square mile of inhabited Earth has more significance for man's future than all the planets in our solar system. It is not the outermost reaches of space, but the innermost recesses of the human soul that now demand our most intense exploration and cultivation. Space exploration, realistically appraised, is only a sophisticated effort to escape from human realities, promoted by Pyramid Age minds, utilizing our advanced Nuclear Age technology, in order to fulfill their still adolescent—or more correctly infantile—fantasies of exercising absolute power over nature and mankind.[24]

Those who insist that a perfect world exists in outer space should remind themselves that without the hatred and antagonism of the Cold War, there would never have been a lunar landing.

The Apollo mission represents the high point of America's love affair with science and technology. The shock of Sputnik frightened Americans into a complete overhaul of their educational system. Billions were injected into the teaching of science and the expansion of universities. Yet by the end of the 1960s, scientists with Ph.D.s were driving taxis or drawing welfare. Universities expanded at a faster rate than they could manage, turning into impersonal education factories that alienated the young. By the end of the decade, the United States had millions of graduates who could solve a quadratic equation or deconstruct Herman Hesse, but not enough plumbers, carpenters, or electricians. Because of the war, the lunar mission, and the Great Society, the country was on the verge of bankruptcy. Americans had demonstrated an ability to build the sophisticated machinery necessary to take them to the Moon, but they could no longer build decent cars. The Pinto exploded and the Gremlin simply broke apart. Americans arrived at parties to celebrate the lunar landing in Toyotas, Datsuns, Volkswagens, and Renaults.

In an April 1964 issue of the *Saturday Evening Post*, Dwight Eisenhower questioned the need for Americans to demonstrate their worth with meaningless stunts in space. Referring to Apollo, he wrote: "This swollen program, costing more than the development of the atomic bomb, not only is contributing to an unbalanced budget; it also has diverted a disproportionate share of our brain-power and research facilities from other equally significant problems, including education." Style, he regretted, had overwhelmed substance; form had smothered content. A people addicted to showmanship would henceforth find it difficult to appreciate subtle but sublime value. There was no going back. One day, Eisenhower feared, historians would judge that "here was where the U.S., like Rome, went wrong—here at the peak of its power and prosperity when it forgot those ideals which made it great."[25] What Eisenhower didn't perhaps realize was that the contagion he bemoaned would quickly spread across the globe as people in other nations eagerly imitated the American taste for artifice. The world might have quickly grown tired of the Moon, a passing fad not unlike the enthusiasm for a rock band or movie star, but it would never cure itself of the addiction to shallow gestures. The Moon has been forgotten, but the madness it encouraged lives on.

A few months after the *Apollo 11* landing on the Moon, the theologian Daniel Migliore ruminated upon the lunar quest in the pages of *Theology Today.* "Space exploration," he said, "seems to offer its own brand of symbolic immortality. The reach of man into outer space is seen by some as a basic drive of the species toward a yet uncertain destiny within the cosmos." This, he felt, arose from the tendency of Americans to confuse progress with movement. Spiritual renewal, they believe, can be achieved simply by moving from Point A to Point B. "Americans are accustomed to transcending the old and the oppressive by spatial movement." He was not opposed to space exploration per se, but he did think it foolish to believe that a perfect world could be discovered somewhere out there.

> Apollo and all that it symbolizes as the American commitment is no future at all if it is a way to evade honest and creative confrontation with our problems of racism, urban blight, militarism, pollution of air and water. It need not be this. But we must make up our minds. Is freedom primarily to be found in spatial migration (to outer or inner space)? Or is freedom to be sought first in shaping a new future for men here on the good earth?[26]

Norman Cousins once said that "What was most significant about the lunar voyage was not that men set foot on the moon, but that they set eye on the Earth."[27] In other words, it took a 240,000-mile journey for men to discover how stunted is their imagination. If there is a lesson to be learned, it is in the futility of seeking fulfillment in outer space. We need to judge ourselves by who we are, not by where we go. The Moon voyage was the ultimate ego trip. Hubris took America to the Moon, a barren, soulless place where humans do not belong and cannot flourish. If the voyage has had any positive benefit at all, it has reminded us that everything that is good resides on Earth.

Notes

PREFACE

1. Diamond, "Dark Side," p. 13.

CHAPTER ONE

1. Burrows, p. 22.
2. Hale, pp. 67, 80.
3. Dickson, p. 31.
4. *Washington Post,* 26 July 1969. See page 240.
5. Benjamin, pp. 18–19.
6. Ehricke, "The anthropology of astronautics," pp. 26–7.
7. See Ehricke, "The extraterrestrial imperative," pp. 18–26.
8. Lewis, *Silent Planet,* 63.
9. Burrows, p. 69.
10. Crouch, p. 24.
11. McCurdy, p. 13.
12. McDougall, p. 18.
13. McCurdy, p. 24.
14. Dickson, p. 43.
15. Pellegrino, p. 181.
16. Ibid., p. 44.
17. Crouch, p. 70.
18. Dickson, p. 45.
19. Carter, p. 94.
20. Titus, p. 77.

CHAPTER TWO

1. Dickson, p. 47.
2. Burrows, p. 52.
3. McCurdy, p. 23.
4. Burrows, p. 77.
5. Ibid., p. 79.
6. Piszkiewicz, pp. 25, 32.

7. Heppenheimer, *Countdown*, p. 15.
8. Piszkiewicz, p. 47.
9. Ibid., p. 65.
10. Dickson, p. 49.
11. Burrows, pp. 99–100.
12. Piszkiewicz, p. 82.
13. Ibid., p. 125.
14. Dornberger, p. 110.
15. Crouch, p. 84.
16. Speer, p. 366.
17. Piszkiewicz, p. 174.
18. Ibid., p. 97.
19. Reuters, 3 April 1995.
20. Piszkiewicz, pp. 136–37.
21. Reuters, 3 April 1995.
22. Piszkiewicz, p. 204.
23. Reuters, 3 April 1995.
24. Hunt, p. 23.
25. Piszkiewicz, pp. 139–40.
26. Michel, p. 99.
27. McDougall, p. 44.
28. Burrows, p. 109.
29. Piszkiewicz, p. 224.
30. Burrows, p. 118.
31. War Department Press Release, 1 October 1945, Internet.
32. Burrows, p. 118.
33. Dickson, pp. 61–62.
34. Hunt, p. 17.
35. *New York Times,* 11 March 1985.
36. Hunt, pp. 18, 20.
37. Memo of the Advisory Committee on Human Radiation Experiments, April 5, 1995, Internet.
38. *New York Times,* 11 March 1985.
39. Hunt, p. 18.
40. Ibid., pp. 19–20.
41. Memo of the Advisory Committee on Human Radiation Experiments
42. Burrows, p. 527.

CHAPTER THREE

1. McDougall, pp. 44–45.
2. Heppenheimer, *Countdown*, p. 19.

3. Holloway, p. 247.

4. CNN, *The Cold War,* 1998, Internet.

5. *New Times,* 10 December 1947.

6. RAND, "Preliminary Design of an Experimental World-Circling Space-ship," May 2, 1946.

7. Ibid.

8. Lipp, np.

9. Perry, np.

10. McDougall, p. 103.

11. Kecskemeti, np.

12. Ibid.

13. McCurdy, p. 36.

14. Ley, *Rockets,* p. 3.

15. Clarke, p. 182.

16. McCurdy, p. 29.

17. Burrows, p. 143.

18. Schefter, p. 14.

19. McDougall, p. 100.

20. Dickson, p. 78.

21. Schefter, p. 14.

22. McCurdy, p. 43.

23. Dickson, p. 79.

24. *New York Times,* 22 June 1955.

25. McDougall, p. 454.

CHAPTER FOUR

1. George Pal, director, *Destination Moon* (1950).

2. McDougall, p. 126.

3. Eisenhower, "The Chance of Peace," Internet.

4. McDougall, p. 119.

5. NSC-5520, 5 May 1955, Eisenhower Library.

6. NSC-5522, 8 June 1955, Eisenhower Library.

7. NSC-5520.

8. McDougall, p. 127.

9. Perry, np.

10. Dickson, p. 86.

11. Launius and McCurdy, p. 29.

12. Dickson, p. 82.

13. Stuhlinger, p. 237.

14. Interview with Maxime Faget, NHO, Folder 16752.

15. Khrushchev, pp. 46–47.

16. Dickson, p. 12.
17. *Washington Post,* 31 August, 9 October 1957.
18. Riabchikov, p. 146.
19. Dickson, p. 13.
20. *New York Times,* 8 November 1957.
21. Pellegrino, p. 3.

CHAPTER FIVE

1. *New York Times,* 17 November 1957.
2. *New York Times,* 15 February 1958; NHO, Folder 6718.
3. "The Impact of Sputnik I," NHO, Folder 6719.
4. *New York Times,* 6 October 1957.
5. *New York Times,* 7 October 1957.
6. *Life,* 21 October 1957.
7. McDougall, p. 301.
8. *Evening Star,* 7 October 1957.
9. Logsdon, pp. 14–16.
10. Dickson, pp. 1, 24.
11. Bergaust, p. 265.
12. Dickson, pp. 252–53.
13. Gunther, pp. 306, 309.
14. Dickson, pp. 134–35.
15. Ibid., p. 15.
16. Medaris, p. 155.
17. McCurdy, p. 62.
18. *New York Times,* 15 October 1957.
19. *Evening Star,* 5 October 1957.
20. *New York Times,* 5 October 1957.
21. Carter, p. 120.
22. *Washington Sunday Star,* 28 December 1958.
23. Dickson, p. 112.
24. *Life,* 21 October 1957.
25. *Washington Post,* 5 October 1957.
26. *Evening Star,* 8 October 1957.
27. Dickson, p. 108.
28. Ibid., pp. 34, 113, 116–17.
29. *Daily News,* 5 October 1957.
30. Dickson, p. 18.
31. *New York Herald Tribune,* 8 October 1957.
32. McCurdy, pp. 74–75.
33. Logsdon, p. 22.

34. McDougall, p. 149.

35. Wolfe, p. 67.

36. Kearns, p. 145.

37. McDougall, pp. 152–53.

38. Ibid., p. 155.

39. Wolfe, p. 67.

40. McDougall, pp. 8, 62.

41. McCurdy, p. 75.

42. McDougall, pp. 221–22.

43. Dickson, p. 224.

44. DeGroot, p. 216

45. Dickson, p. 139.

46. *New York Herald Tribune,* 16 October 1957.

47. *Missiles and Rockets,* November 1957.

48. Billy Graham to Eisenhower, 2 December 1957, NHO, Folder 15322.

49. Burrows, p. 186.

50. *Wall Street Journal,* 11 February 1960.

51. Dickson, pp. 226, 229.

52. Ibid., pp. 122, 126.

53. McDougall, p. 158.

54. Dickson, p. 139.

55. *Washington Post,* 8 October 1957.

56. *Washington Post,* 10 October 1957.

57. Crouch, p. 92.

58. Dulles to Hagerty, 8 October 1957, Eisenhower Library.

59. McAleer, p. 13.

60. Dickson, p. 223.

CHAPTER SIX

1. *New York Times,* 16 May 2000; *Nature,* 4 May 2000.

2. NHO, Folder 16752.

3. *Washington Post,* 15 October 1957.

4. *Air and Space,* October/November 1990, p. 112.

5. Swanson, p. 64.

6. *Missiles and Rockets,* November 1957.

7. Dickson, p. 147.

8. Schefter, p. 27.

9. Dickson, p. 156.

10. Wolfe, p. 68.

11. Dickson, pp. 158–59.

12. Richelson, np.

13. Manchester, p. 66.
14. Richelson, np.
15. Ibid.
16. Ibid.
17. Richelson, McCurdy, p. 76.
18. Dickson, p. 123.
19. Ibid., p. 175.
20. Schefter, p. 33.
21. Dickson, p. 176.
22. Ibid., p. 125.
23. Schefter, pp. 34–36.
24. Dickson, p. 125.
25. Schefter, pp. 34–36.
26. Heppenheimer, *Countdown*, p. 115
27. Pellegrino, p. 37.
28. Dickson, p. 177.
29. DeGroot, p. 256.
30. Dickson, p. 178.
31. Ibid., p. 186.
32. Both letters are published in *Christian Century*, 23 December 1959.
33. Dickson, p. 186.
34. Launius and McCurdy, p. 33.
35. Logsdon, p. 26.
36. McDougall, p. 182.
37. Swanson, p. 67.
38. Dickson, p. 193.
39. Ibid., p. 201.
40. Ibid., pp. 200–201.
41. McDougall, p. 204.
42. Schefter, p. 112.
43. *New York Times,* 4 November 1957.
44. See, for instance, United States Information Agency, "The impact of Sputnik on the standing of the U.S. versus the USSR" (December 1957), and "Post-Sputnik attitudes to NATO and Western Defense" (February 1958), NHO, Folder 6737.

CHAPTER SEVEN

1. Burrows, p. 215.
2. McCurdy, p. 48.
3. NASA Long Range Plan, 1959, NHO.
4. Burrows, p. 267.

5. Walsh, p. 18.

6. Scott and Leonov, p. 68.

7. Crouch, p. 156.

8. Glenn, *Memoir,* pp. 186, 197.

9. J. W. Humphreys to Nicolae Craciunescu, 1 December 1969, NHO, Folder 6618.

10. Collins, *Carrying the Fire,* p. 39.

11. Wolfe, p. 101.

12. Scott and Leonov, p. 278.

13. Burrows, pp. 298–300.

14. Carpenter, p. 11.

15. *Public Papers of the Presidents: Lyndon Johnson 1963–64,* p. 1464.

16. Glenn, *Memoir,* p. 197.

17. Schefter, p. 63.

18. Ibid., p. 64–65, 68.

19. Wolfe, p. 107.

20. *Public Papers of the Presidents: John F. Kennedy 1962,* p. 159.

21. Kauffman, p. 34.

22. *New York Times,* 27 May 1962.

23. *Newsweek,* 5 March 1962, 27 May 1963.

24. Burrows, p. 295.

25. Slayton, p. 185

26. Schefter, pp. 86–87.

27. Ibid., p. 67.

28. Dora Jane Hamblin to P. Michael Whye, 18 January 1977, NHO, Folder 6719.

29. *Life,* 14, 21 September 1959.

30. Kauffman, p. 74.

31. Hamblin to Whye, 18 January 1977.

32. Scott and Leonov, pp. 85–86.

33. Diamond, "The *Life* contract," p. 17.

34. Kauffman, p. 55.

35. Swanson, p. 25.

36. Schefter, p. 72.

37. Ibid., p. 74.

38. Burrows, p. 300.

39. Hamblin to Whye, 18 January 1977.

40. *Denver Post,* 17 July 1969.

41. Heppenheimer, *Countdown,* p. 157.

42. Swanson, pp. 313–14.

43. Ibid., p. 316.

44. Dickson, p. 215.

45. Transcript of Presidential Debate, 21 October 1960, JFK Library.

46. Launius and McCurdy, p. 52.

47. *Oregonian*, 8 September 1960.

CHAPTER EIGHT

1. Kennedy Inaugural Address, 20 January 1961, JFK Library.

2. Schefter, p. 112.

3. Hornig committee report, NHO.

4. Ibid.

5. Logsdon, p. 35.

6. McDougall, p. 225.

7. Eisenhower Farewell Address, 17 January 1961, Eisenhower Library.

8. McDougall, p. 230.

9. Weisner Committee Report, NHO.

10. Duff interview, NHO.

11. *Time*, 4 October 1963.

12. Kauffman, p. 14.

13. JFK News Conference, 8 February 1961, JFK Library.

14. McDougall, p. 317.

15. Excerpts from Berkner Report, "From the Chair," *Space Studies Bulletin*, January–March 2003, p. 2.

16. Logsdon, p. 89.

17. Burrows, p. 294.

18. Schefter, p. 136.

19. Pellegrino, p. 21.

20. Carter, p. 157.

21. *Washington Post*, 13 April 1961.

22. *New York Times*, 16 April 1961.

23. *Congressional Record*, 24 May 1962.

24. *Washington Post*, 7 May 1962.

25. *New York Times*, 23, 26 January 1959.

26. McDougall, pp. 246–47.

27. Kennedy Press Conference, 12 April 1961, JFK Library.

28. Sidey, pp. 121–23.

29. Kauffman, p. 102.

30. Von Bencke, p. 67.

31. Logsdon, p. 103.

32. White House Meeting, 21 November 1962, Presidential Recording Tape 63, JFK Library.

33. Kennedy Press Conference, 21 April 1961, JFK Library.

34. Kennedy to Johnson, 20 April 1961, LBJ Library.

35. WAMU, Wilson interview.

36. McDougall, p. 319.

37. Ibid., p. 322.

38. Johnson to Kennedy, 28 April 1961, LBJ Library.

39. Wolfe, p. 213; Burrows, p. 326.

40. Duff interview, NHO.

41. Logsdon, p. 125.

42. Swanson, p. 80.

43. Swanson, pp. 10, 68.

44. Kennedy Speech, 25 May 1961, JFK Library.

45. Schefter, p. 143.

46. Burrows, p. 329.

47. Alexander, p. 117.

48. Kennedy Inaugural Address, JFK Archives.

49. 25 May 1961, Speech, JFK Library.

50. Launius and McCurdy, p. 58.

51. Wheelon and Graybeal, p. 1.

52. Kauffman, p. 33.

53. McDougall, p. 388.

54. JFK Press Conference, 10 August 1961, JFK Library.

55. Kauffman, p. 127.

56. *Evening Star,* 25 April 1963.

57. *Congressional Record,* 10 October 1963.

58. Kauffman, p. 12.

59. Memo, Webb to Kennedy, 21 March 1961, State Department Archives.

60. Kauffman, pp. 21, 66.

61. Manchester, p. 66.

62. Launius and McCurdy, p. 62.

63. *Saturday Evening Post,* 11 August 1962, p. 24.

64. McDougall, p. 394.

65. *Space Daily,* 26 June 1963.

66. Kauffman, p. 22.

67. Jastrow and Newell, pp. 42–43.

68. Rice University Speech, 12 September 1962, JFK Library.

69. Kauffman, p. 25

70. Parker, p. 306.

71. Kauffman, p. 27.

72. *Wall Street Journal,* 12 September 1962.

73. Webb, p. 75.

74. *Wall Street Journal,* 9 July 1969.

75. *New York Times,* 20 July 1960; Stanford Research Institute, "Benefits from Space," March 1969, p. 9, NHO, Folder 6911.

76. Etzioni WAMU interview.
77. McDougall, p. 438.
78. *San Francisco Sunday Examiner and Chronicle,* 17 March 1968.
79. McDougall, p. 384.
80. Kauffman, pp. 24, 98–100.
81. *Massapequa Post,* 1 May 1963.
82. *New York Times,* 3 May 1963.
83. Kauffman, p. 123.
84. McDougall, p. 391.
85. *Washington Post,* 9 April 1963.
86. Logsdon, pp. 107, 110–11.
87. Kauffman, p. 102.

CHAPTER NINE

1. Viorst, p. 68.
2. Schefter, p. 175.
3. Pellegrino, pp. 30–31.
4. McCurdy, p. 168.
5. Pellegrino, pp. 30–31.
6. Kennedy Press Conference, 11 October 1961, JFK Library.
7. Schefter, p. 180.
8. Ibid., pp. 154–56.
9. Kennedy Press conference, 7 February 1962.
10. Kauffman, pp. 64–65.
11. McCurdy, p. 216.
12. Rio de Francisco, "The John Glenn Twist," 1962.
13. *New York Times,* 9 March 1962.
14. *Life,* 9 March 1962.
15. *Congressional Record,* 26 February 1962.
16. *Catholic Standard,* 9 March 1962; *New York Times,* 1 March 1962.
17. *Washington Post,* 7 March 1963.
18. Kennedy Press Conference, 22 February 1962, JFK Library.
19. Ibid.
20. State of the Union Address, 30 January 1961, JFK Library.
21. Kennedy Speech at University of California, 23 March 1962, JFK Library.
22. Kennedy Address to the UN, 23 September, 1963, JFK Library.
23. McDougall, p. 350.
24. Schefter, p. 172.
25. Ibid., p. 181.
26. Kennedy Speech at Rice University, 12 September 1962, NHO.

27. Swanson, pp. 7–8.

28. White House Meeting, 21 November 1962, Presidential Recording Tape 63, JFK Library.

29. Ibid.

30. Webb to Kennedy, 30 November 1962, NHO.

31. *Los Angeles Herald-Examiner,* 15 May 1963; *New York Times,* 22 May 1963.

32. Heppenheimer, *Countdown,* p. 215.

33. Burrows, p. 347.

34. Schefter, p. 192.

35. Levine, pp. 146–47.

36. *Evening Star,* 6 July 1963.

37. Ackmann, pp. 167, 173.

38. Scott and Leonov, p. 77.

39. Alien Plant Working Group Web site.

40. Crouch, p. 175.

41. Swanson, p. 82.

42. Launius and McCurdy, p. 73.

43. Internal Memo, 19 September 1967, NHO, Folder 7039.

44. WAMU, Etzioni interview.

45. Kauffman, p. 50.

46. Hogan, p. 21.

47. Swanson, p. 11

48. Duff interview, NHO.

49. *Congressional Record,* 18 January 1962.

50. *Washington Post,* 21 June 1963.

51. Ibid.

52. "Washington Goes to the Moon," Part 1.

53. *Washington Post,* 11 June 1963.

54. McCurdy, p. 217.

55. *Saturday Review,* 4 August 1962.

56. *The Virginian-Pilot,* 3, 4 April 1963.

57. *San Francisco Chronicle,* 29 May 1963.

58. Leonov and Scott, p. 80.

59. Kauffman, p. 14.

60. Webb, Address to Institute of Foreign Affairs, 24 January 1963, NHO.

61. *Birmingham Post-Herald,* 25 April 1963.

62. *Life,* 17 May 1963.

63. *Washington Post,* 7 May 1963.

64. *Birmingham News,* 5 March 1969.

65. Kauffman, pp. 117–18.

66. *Daily Press,* 2 June 1963.

67. Diamond, *Rise and Fall,* p. 1ff.

68. "Washington Goes to the Moon," Part 1.

69. Ibid.

70. WAMU, Etzioni interview.

71. Kennedy Press Conference, 22 August 1963, JFK Library.

72. *New York Times*, 27 October 1963.

73. McDougall, p. 291.

74. Kennedy Press Conference, 31 October 1963, JFK Library.

75. Remarks Intended for Delivery to the Texas Democratic State Committee in the Municipal Auditorium in Austin, 22 November 1963, JFK Library.

76. Scott and Leonov, p. 89.

CHAPTER TEN

1. Scheer memo, 6 May 1965, NHO, Folder 6414.

2. NHO Folder 6414.

3. *Washington News*, 8 March 1972.

4. Gary Kern to John F. Kennedy, NHO, Folder 6060.

5. See "Models," NHO, Folder 6412.

6. "NASA Indicators," NHO, Folder 6716.

7. *Washington Post*, 9 April 1970.

8. *Aviation Week and Space Technology*, 1 May 1972.

9. Duff interview, NHO.

10. Pellegrino, p. 45.

11. Swanson, p. 213.

12. NHO Folder 6718.

13. Launius, "Evolving Perceptions," pp. 825–26.

14. Heppenheimer, *Countdown*, p. 217.

15. Crouch, p. 194.

16. Scott and Leonov, p. 134.

17. Cunningham, pp. 211, 221.

18. Scott and Leonov, p. 107.

19. Walsh, p. 47.

20. Scott and Leonov, p. 168.

21. Schefter, p. 236.

22. Scott and Leonov, p. 169.

23. Schefter, p. 238.

24. WAMU, Seamans interview.

25. Scott and Leonov, p. 170.

26. Ibid., p. 170.

27. WAMU, Seamans interview.

28. Scott and Leonov, pp. 180–81.

29. Ibid., p. 159.

30. Cernan, p. 133.

31. McCurdy, p. 99.

32. Biddle, pp. 32–33.

33. "Washington Goes to the Moon," Part 1.

34. WAMU, Seamans interview.

35. Johnson, *The Vantage Point,* p. 285.

36. *New York Times,* 14 August 1966.

37. WAMU, Schultze interview.

38. Launius and McCurdy, p. 82.

39. Benjamin, p. 10.

40. Swanson, pp. 219–20.

41. Schefter, p. 242.

42. Swanson, pp. 216, 219.

43. Burrows, p. 475.

CHAPTER ELEVEN

1. Pellegrino, p.112.

2. Scott and Leonov, p. 184.

3. Ibid., pp. 184–85.

4. Ibid., pp. 185–86.

5. Swanson, p. 122.

6. Scott and Leonov, pp. 186–87.

7. Swanson, pp. 122–23.

8. *Washington Post,* 5 April 1962.

9. Walsh, pp. 60–61.

10. WAMU, Seamans interview.

11. Kraft, p. 274.

12. Swanson, pp. 124–26.

13. *Life,* 3 February 1967.

14. *Boston Globe,* 12 April 1967.

15. WAMU, Schultze interview.

16. Alexander, p. 115.

17. Phillips Report, NHO.

18. "Washington Goes to the Moon," Part 2.

19. Pellegrino, pp. 89, 106, 111–12.

20. *Wall Street Journal,* 20 February 1967.

21. WAMU, Mondale interview.

22. Senate Apollo Hearings, NHO.

23. WAMU, Seamans interview.

24. Senate Apollo Hearings, NHO.

25. Mueller to Atwood, 19 December 1965, NHO.

26. "Washington Goes to the Moon," Part 2.
27. Ibid.
28. WAMU, Mondale interview.
29. WAMU, Seamans interview.
30. WAMU, Mondale interview.
31. WAMU, Seamans interview.
32. "Washington Goes to the Moon," Part 2.
33. Turnill, p. 211.
34. Ibid.
35. Scott and Leonov, p. 192.
36. Swanson, pp. 372–74.
37. Burrows, p. 410.
38. Alexander, p. 153.
39. Pellegrino, pp. 131–32.
40. Ibid., p. 133; Biddle, p. 37.
41. Schefter, pp. 243–44.
42. Pellegrino, p. 137.

CHAPTER TWELVE

1. Schefter, p. 253.
2. "Washington Goes to the Moon," Part 2.
3. Pellegrino, p. 122.
4. Swanson, p. 111.
5. Ibid., p. 122.
6. Schefter, p. 249.
7. WAMU, Seamans interview.
8. Pellegrino, p. 163.
9. Turnill, p. 139.
10. Swanson, p. 36.
11. *Washington Post*, 31 July 1967.
12. Kranz, p. 232.
13. Burrows, p. 418.
14. Kraft, p. 291.
15. Swanson, p. 130.
16. Pellegrino, p. 167.
17. Burrows, p. 419.
18. Turnill, p. 171.
19. Swanson, pp. 134–35.
20. Ibid., p. 214
21. Zimmerman, p. 232.
22. Pellegrino, p. 184.

23. Lewis, "Onward Christian Spacemen," p. 29.
24. Benjamin, p. 47.
25. McCurdy, p. 99.
26. Turnill, p. 235.
27. *Washington Post*, 26 March 1969.
28. *New York Times*, 25 December 1968.
29. Benjamin, p. 49.
30. Scott and Leonov, p. 238.
31. Ibid., pp. 240–41.
32. Swanson, p. 105.

CHAPTER THIRTEEN

1. Mangan, p. 4.
2. Press Conference, 5 July 1969, NHO.
3. *Newsweek*, 7 July 1969.
4. Heppenheimer, *The Space Shuttle Decision*, np.
5. Migliore, p. 439.
6. Hogan, p. 26.
7. Swanson, p. 160.
8. Ibid., pp. 160–61.
9. Ibid., p. 255.
10. Turnill, p. 259.
11. Scott and Leonov, p. 246.
12. Diamond, "Dark Side," pp. 13, 16.
13. Ibid., pp. 13, 16.
14. Ibid., pp. 14, 16.
15. *Baltimore Sun*, 30 July 1969.
16. Scott and Leonov, pp. 247–48.
17. Carter, pp. 197–98, 207–8.
18. My thanks to Michael Fisher.
19. *Boston Sunday Globe*, 7 September 1969.
20. *TV Guide*, 19 July 1969.
21. Mailer, p. 70.
22. Benjamin, p. 57.
23. *Washington Post*, 26 July 1969.
24. Benjamin, p. 61.
25. Walsh, p. 95.
26. Collins, *Liftoff*, p. 7.
27. Scott and Leonov, p. 291.
28. Hogan, p. 35.
29. Swanson, p. xii.

30. *New York Times,* 17 July 1969.

31. Turnill, pp. 221–22.

32. "Statement by President Nixon on the Space Program," 7 March 1970, NHO.

33. Heppenheimer, *Shuttle,* np.

34. Ibid.

35. Ehrlichman, pp. 144–45.

36. Report of the Space Task Group, 1969, NHO.

37. Heppenheimer, *Shuttle,* np.

38. Ibid.

39. See NHO, Folder 6716, "Public Opinion 1967–69."

40. Heppenheimer, *Shuttle,* np.

41. Ibid.

42. Congressional Record, July 6, 1970.

43. Heppenheimer, *Shuttle,* np.

44. Ibid.

45. Walsh, p. 99.

46. Ibid.

47. *All in the Family,* CBS Television, 1972.

48. *Washington Star,* 8 May 1957.

49. Scott and Leonov, p. 301.

50. *New York Times,* 11 December 1972.

51. Cernan, p. 340.

52. WAMU Schultze interview.

53. WAMU Cronkite interview.

54. *New York Times,* 22 July 1974.

55. Pellegrino, p. 284.

CHAPTER FOURTEEN

1. McCurdy, p. 202.

2. *Daily News,* 11 March 1979.

3. Mailer, pp. 30, 41.

4. Swanson, p. 217.

5. Mailer, p. 43.

6. Benjamin, p. 57–59.

7. *Houston Chronicle,* 19 July 1970.

8. Carter, p. 199.

9. McCurdy, p. 241.

10. Hogan, p. 12.

11. Heppenheimer, *Shuttle,* np.

12. Pellegrino, p. 282.

13. Webb, p. 8.

14. Declaration for Space, 1986, NHO.

15. Carter, p. 200.

16. Benjamin, p. 20.

17. McCurdy, p. 151.

18. Benjamin, p. 127.

19. Ibid., p. 131.

20. Ibid., pp. 131–32.

21. http://members.aol.com/oscarcombs/wannabe.htm.

22. George Mueller, "On the 25th Anniversary of Apollo," text of speech given on 20 July 1994, NHO, Folder 6723.

23. Jim Dator, "Politics is Everything," Space and Society Lecture, Strasbourg, 4 December 1995.

24. *Newsweek*, 7 July 1969.

25. *Saturday Evening Post*, 11 April 1964, p. 18.

26. Migliore, p. 445.

27. Scott and Leonov, p. 378.

Bibliography

ARCHIVE COLLECTIONS

Dwight D. Eisenhower Library
Lyndon B. Johnson Library
John F. Kennedy Library
Richard M. Nixon Library
Harry S. Truman Library

National Archives, Washington
NASA History Office
U.S. Department of State

Johnson Space Center
Kennedy Space Center

BOOKS AND ARTICLES

(ALL BOOKS PUBLISHED IN NEW YORK UNLESS OTHERWISE NOTED)

Ackmann, Martha. *The Mercury 13.* (2003)
Alexander, Tom. "The unexpected payoff of Project Apollo," *Fortune,* July 1969.
Benjamin, Marina. *Rocket Dreams.* (2003)
Bergaust, Erik. *Wernher von Braun.* (Washington, D.C., 1976)
Biddle, Wayne. "A great new enterprise," *Air and Space Smithsonian,* June/July 1989.
Burrows, William E. *This New Ocean.* (1999)
Bush, Vannevar. *Modern Arms and Free Men.* (1949)
Carpenter, Scott, et al. *We Seven.* (1962)
Carter, Dale. *The Final Frontier.* (London, 1988)
Cernan, Eugene. *The Last Man on the Moon.* (1999)
Chaikin, Andrew. *A Man on the Moon.* (1998)
Clarke, Arthur. *Exploration of Space.* (1951)
Collins, Michael. *Carrying the Fire.* (2001)
Collins, Michael. *Liftoff: The Story of America's Adventure in Space.* (1988)
Crouch, Tom D. *Aiming for the Stars.* (Washington, D.C., 1999)

Cunningham, Walter. *The All-American Boys.* (2003)

DeGroot, Gerard. *The Bomb: A Life.* (London, 2004)

Diamond, Edwin. "The dark side of the Moon coverage," *Columbia Journalism Review,* Fall, 1969.

Diamond, Edwin. "The *Life* contract with NASA: What price exclusivity?" *Columbia Journalism Review,* Fall, 1969.

Diamond, Edwin. *The Rise and Fall of the Space Age.* (1964)

Dickson, Paul. *Sputnik.* (2001)

Dornberger, Karl. *V-2.* (1979)

Ehricke, Krafft. "The anthropology of astronautics," *Astronautics,* November 1957.

Ehricke, Kraft. "The extraterrestrial imperative," *Bulletin of the Atomic Scientists,* November 1971.

Ehrlichman, John. *Witness to Power.* (1982)

Etzioni, Amitai. *Moon-Doggle.* (1964)

Glenn, John. *A Memoir.* (1999)

Greenberg, D. S. "Civilian technology: NASA study finds little 'spin-off,'" *Science,* 1 September 1967.

Gunther, John. *Inside Russia Today.* (1962)

Hale, Edward. *The Brick Moon and Other Stories.* (1869)

Harland, David. *Exploring the Moon.* (Chichester, U.K., 1999)

Harrison, Albert. *Spacefaring.* (Berkeley, 2001)

Heppenheimer, T. A. *Countdown.* (1997)

Heppenheimer, T. A. *The Space Shuttle Decision.* NHO. (1999)

Hogan, Alfred. "Science on the set," unpublished thesis, University of Maryland. (1986)

Holloway, David. *Stalin and the Bomb.* (New Haven, 1994)

Hunt, Linda. "US coverup of Nazi scientists," *Bulletin of the Atomic Scientists,* April 1985.

Jastrow, Robert, and Homer Newell. "Why land on the Moon?," *The Atlantic,* August 1963.

Johnson, Lyndon. *The Vantage Point.* (1971)

Johnson, Stewart, et al., eds. *Space 2000.* (Albuquerque, 2000)

Kauffman, James L. *Selling Outer Space.* (Tuscaloosa, 1994)

Kearns, Doris. *Lyndon Johnson and the American Dream.* (1976)

Kelly, Thomas. *Moon Lander.* (Washington, D.C., 2001)

Kecskemeti, Paul. "The satellite rocket vehicle: Political and psychological problems," RAND, 4 October 1950.

Khrushchev, N. S. *Khrushchev Remembers: The Last Testament.* (1974)

Klerkx, Greg. *Lost in Space.* (London, 2004)

Kraft, Chris. *Flight: My Life in Mission Control.* (2001)

Kranz, Gene. *Failure Is Not an Option.* (2000)

Launius, Roger. "Evolving perceptions of spaceflight in American culture," *Acta Astronautica* (2003)

Launius, Roger, and Howard McCurdy, eds. *Spaceflight and the Myth of Presidential Leadership.* (Urbana, Ill., 1997)

Lewis, C. S. "Onward Christian spacemen," *Show,* February 1963.

Lewis, C. S. *Out of the Silent Planet.* (2003)

Levine, Sol. *Appointment in the Sky.* (1963)

Ley, Willy. *Rockets, Missiles and Space Travel.* (1944)

Ley, Willy. *The Conquest of Space.* (1949)

Lipp, J. E. "Reference papers relating to a satellite study," RAND Report RA-15032, 1 February 1947.

Logsdon, John. *The Decision to Go to the Moon.* (Chicago, 1970)

McAleer, Neil. "The space age turns 30," *Space World,* October 1987.

McCurdy, Howard. *Space and the American Imagination.* (Washington, D.C., 1997)

McDougall, Walter A. *The Heavens and the Earth.* (Baltimore, 1985)

Mailer, Norman. *A Fire on the Moon.* (1970)

Manchester, Harland. "The senseless race to put man in space," *Reader's Digest,* May 1961.

Mangan, Terence. "The Doman Moon Chapel," *Liturgical Arts,* November 1967.

Medaris, John. *Countdown for Decision.* (1960)

Michel, Jean. *Dora.* (1979)

Migliore, Daniel. "Theological table-talk," *Theology Today,* January 1970.

Moore, Patrick. *New Guide to the Moon.* (1976)

Parker, P. J. "Benefits of space," *Spaceflight,* September 1968.

Pellegrino, Joseph, and Joshua Stoff. *Chariots for Apollo.* (1999)

Perry, Robert L. *Origins of the USAF Space Program.* (1961)

Piskiewicz, Dennis. *The Nazi Rocketeers.* (Westport, Conn., 1995)

RAND. "Preliminary design of an experimental world-circling spaceship," 2 May 1946.

Read, David. "Sputnik and the angels," *Christianity Today,* 9 December 1957.

Riabchikov, Evgeny. *Russians in Space.* (1971)

Richelson, Jeffrey. "Shooting for the Moon," *Bulletin of the Atomic Scientists,* September/October 2000.

Schefter, James. *The Race.* (1999)

Scott, David, and Alexei Leonov. *Two Sides of the Moon.* (London, 2004)

Sidey, Hugh. *John F. Kennedy, President.* (1964)

Slayton, Donald. *Deke!* (1994)

Smith, Andrew. *Moondust.* (London, 2005)

Speer, Albert. *Inside the Third Reich.* (1977)

Stuhlinger, Ernst. "Sputnik 1957—memories of an old-timer," *JBIS,* July/August 1999.

Swanson, Glen E., ed. *"Before This Decade Is Out . . . ".* (Gainesville, 2002)

Titus, A. Constandina. *Bombs in the Backyard*. (Reno, 1986)

Turnill, Reginald. *The Moonlandings*. (Cambridge, 2003)

Verne, Jules. *From the Earth to the Moon*. (1967)

Viorst, Milton. *Hustlers and Heroes*. (1971)

Von Bencke, Matthew. *The Politics of Space: A History of US-Soviet/Russian Competition and Cooperation in Space*. (1997)

Walsh, Patrick. *Echoes among the Stars*. (Armonk, N.Y., 2000)

Webb, James. "Commercial use of space research and technology," *Astronautics and Aerospace Engineering*, June 1964.

Wheelon, Albert, and Sidney Graybeal. "Intelligence for the space race," *Studies in Intelligence*, Fall, 1961.

Wolfe, Tom. *The Right Stuff*. (London, 1991)

Zimmerman, Robert. *Genesis: The Story of Apollo 8*. (1998)

Index

Page numbers followed by one asterisk signify the first footnote on the page; page numbers followed by two asterisks signify the second footnote on the page.

2018: A Hopeful View of the Human Future (O'Neill), 263–264

Abelson, Philip, 173–174
Abernathy, Ralph, 234–235
ABMA. *See* Army Ballistic Missile Agency (ABMA)
Adams, Sherman, 80
Aerospace industry, 137–138, 202
Agena satellite, 194, 195, 197–198
Agnew, Spiro, 244–245, 246
Albert rocket, 38
Albert II rocket, 38
Aldrin, Buzz: Communion service on the Moon, 236; end of space race, 252; lunar landing, xiii; statement on setting foot on the Moon, 241
Alien Encounter ride, 259
All in the Family (TV show), 251
Allen, Steve, 237
Alvarez, Luis, 262
American Civil Liberties Union, 236
American Mercury (magazine), 64
American Rocket Society, 61, 61*
Anders, William, 227, 229
Anderson, Clinton, 145, 246
Anfuso, Victor, 135

Apollo 1, 205–220; builder, 206; command module test, 208–209; deadline for lunar landing, 208; fire onboard, 208–209, 219–220, 223, 232; fire's aftermath, 212–219, 223–225; flammable materials, 209–210, 219–220; Grissom and, 205–206, 208–209; Kraft and, 209, 210; Kranz on, 207, 210; Leonov on, 219; *Life* (magazine), 211; manned mission, 207; McDivitt and, 206; memorial services for, 220; Mueller and, 223–224; North American Aviation, 206, 217*; pure oxygen environment, 219; Schweickart and, 206; Scott and, 206; Seamans on, 209, 224; service module, 208; Shepard on, 223; White and, 205–206, 208–209
Apollo 6, 224
Apollo 7, 225–226
Apollo 8, 186, 227–230, 236
Apollo 9, 230–232
Apollo 10, 232, 250
Apollo 11, 234–240; American Civil Liberties Union, 236; Armstrong (Neil) and, 235; Cronkite and, 235, 239; Kennedy's (John F.) grave, 181; lunar landing (*see* Lunar landing); media coverage, xi–xii, 237–238; memories of, 250; *New York Times* on, 239; Nixon on, 2, 242–243; Poor People's Campaign,

234; product advertisements, 237; public response, 238–240; Saturn V rocket, 235; Soviet response, 238; take off, 235; television, xi–xii, 239

Apollo 12, 248–249

Apollo 13, 249–251

Apollo 13 (film), 250

Apollo 14, 251

Apollo 15, 241, 251

Apollo 17, 252, 261

Apollo program: American love affair with science and technology, 267; Christmas Eve message from space, 227, 228; command module, 207, 208–209, 227; complexity, 205; cost, 242; deadline for lunar landing, 206–207, 208, 212, 220–221, 224, 232, 242; docking in space, 231; Eisenhower and, 123, 267–268; Hornig committee report (1960), 122; human wreckage, 221; legacy of, 260–261; lunar landing, 122; lunar module, 227; Lunar Rover, 251*; media stunts, 226; Moon package, 224; morale, 223; NASA, 102; *New York Times* on, 200; non-astronaut fatalities, 220–221; orbiting the Moon, 227, 230–231; oxygen tanks, 250; Phillips report, 213–214, 216–218; program manager, 207; Saturn V rocket, 224, 235; schedule, 205–206, 223, 227; sense of security/complacency, 208, 210; software, 227; subcontractors, 206; testing, 207–208, 223–224, 227, 232; "Tiger Team" investigation, 213–214; training, 207; wiring, 220

Arms race: Eisenhower and, 46, 51; satellites, 51; Sputnik, 68

Armstrong, Jan, 195–196, 203, 221–222

Armstrong, Neil: *Apollo 11,* 235; *Declaration for Space,* 262; end of space race, 252; *Gemini 8,* 195–197, 208; lunar walk, xiii, 3; press conferences, 255–256; purpose of lunar mission, 233; small step, 258, 262; space exploration, 257; Spaceship NASA, 247; statement on landing on the Moon, 240–241; walk on the Moon, 204; winged gospel, 2–3

Army Air Corps, 32

Army Ballistic Missile Agency (ABMA): NASA, 96, 101; U. S. Army, 49, 53, 55; von Braun (Wernher) and, 53

Asimov, Isaac, 261

Assignment—Outer Space (film), 183

Astrodome, 185

Astronautics, Three Laws of, 3

Astronauts: Astronaut Hall of Fame, 115; belief in lunar mission, 215; Columbus compared to, 109; control exercised by, 193; cosmonauts compared to, 103; divorce, 192; food, 226; fun-loving guys, 190, 191; gas pains, 226; Johnson on, 144; marketability, 191, 211; media stunts, 191; Mercury astronauts (*see* Mercury astronauts); Mercury compared to Gemini astronauts, 192; *Newsweek* on, 109, 114; novelty value, 192; Order of the Ancient Astronauts, 255; personal items in space, 190–191; public role, 192, 205; rebellion by, 226; Russian (*see* Cosmonauts); scientists/engineers compared to, 191; space groupies, 116, 221; space race, 192–193; split personalities, 191; "squawk boxes," 195; *Time* picture, 176; wannabe astronauts,

185; wives of, 113, 116–117, 195–196, 221–222

Astroturf, 185

Atlantic Monthly (magazine), 1, 148

Atlas rocket: builder, 103; failures, 117–118, 119, 135, 139, 194; Gemini program, 194; Kennedy (John F.) and, 119; thrust, 122

Atmospheric test ban treaty, 175

Atomic power, 10

Atterley, Joseph, 5

Atwood, Lee, 217

Aurora 7 flight, 163

Baer, Steve, 264

Baker, Bobby, 153, 182*

"Ballad of John Glenn, The" (song), 159

Ballistic missiles, 31, 34, 91. *See also* Army Ballistic Missile Agency (ABMA); ICBMs (intercontinental ballistic missiles)

Bardot, Brigitte, 115

Bartholomae, Sara, 160

Bartholomae, William, 160

Baruch, Bernard, 73–74

Bassett, Charles, 197

Bay of Pigs, 135, 147, 200

Bean, Alan, 249

Becker, Karl Emil, 13

Beethoven, Ludwig van, 239

Belka (a dog), 99, 120

Bell, David, 134, 166

Bell Aerosystems, 154

Bell Aircraft, 182*

Belyayev, Pavel, 227

Bennett, Rawson, 80

Bergaust, Erik, 81

Bergerac, Cyrano de, 5

Bergman, Jules, 224

Berkner, Lloyd: International Geophysical Year (IGY), 47–48; Space

Science Board report (1961), 129–130, 137; Sputnik launch, 59; Webb and, 129, 130

Bernal, J. D., 3, 11

Birmingham News (newspaper), 66, 177

Black, Dwight, 85

Blagonravov, A. A., 68

Blagonravov, A. A., 68

"Bless Thou the Astronauts" (song), 184

Boeing, 86

Bogdonov, Alexander, 5

Bondarenko, Valentin, 219, 219*

Bonestell, Chesley, 39, 40

Bonney, Walt, 112

Bordeyev, Mikhail, 155*

Borman, Frank, 176, 194, 227–229

Boston Globe (newspaper), 211

Boushey, Homer, 84–85

Bowie, David, 184

Bradbury, Ray, 237

Branson, Richard, 13

Breaking the Sound Barrier (film), 77**

Brezhnev, Leonid, xiii

"Brick Moon, The" (Hale), 1–2, 263

Bridges, Styles, 73, 95

Brooks, Overton, 135, 171

Bryan, William Jennings, 70

Buck Rogers (fictional character), xiii, 8, 9, 38, 42, 119, 131

Buckley, William F., 177

Bumper 8 rocket, 36–37

Bundy, McGeorge, 142

Burrows, William, 155*

Bush, George H. W., 255

Bush, George W., 255

Bush, Vannevar, 34, 146–147

Bykovsky, Valery, 167

Byrd, Harry, 169

Byrds, The, 184

Cameron, James, 237

Campbell, Robert, 167

Canning, Thomas, 182*
Cannon, Howard, 151
Cape Canaveral (Florida), 36–37, 115, 170
Capitalism, lunar mission and, 237
Capra, Frank, 183
Captain Kirk (fictional character), 260, 262
Carpenter, Rene, 112
Carpenter, Scott: *Aurora 7* flight, 163; car driven by, 115; education, 105–106; interest in astronomy, 106; philosophical tendencies, 106; public face, 107; selection as astronaut, 105–106; on space, 109
Carson, Johnny, 252
Cassini space probes, 94
Castenholz, Paul, 90
Castro, Fidel, 135
Celebi, Lagari Hasan, 1, 229
Cernan, Gene, 197–198, 232, 252
Chaffee, Roger, 205–206, 210, 211
Chamberlain, Neville, 73
Chapman, Sydney, 47
Checkers, 120*
Chicago Daily News (newspaper), 68
Chimpanzees in space: Ape No. 61, 131; Chimp College, 168; Enos, 157, 255; films, 183; Glenda, 168; Ham, 131–132, 157, 255; *Life* magazine cover, 131; Mercury program, 130–132, 157–158; Shepard on, 131
Churchill, Winston, 73
CIA (Central Intelligence Agency): prediction of Sputnik launch date, 57; response to Sputnik, 62, 63; World's Fair (1958), 100
Civil rights movement, 65–66, 199, 234–235
Clark, Joseph, 151
Clarke, Arthur C.: on *Apollo 8*, 230; *Apollo 11*, 237; manned space

flight, 39–40; Spaceship NASA, 247; on Sputnik, 63; von Braun (Wernher) and, 53
Cobb, Geraldyn, 168
Coca-Cola Company, 184
Cocoa Beach (Florida), 115, 221
Cold War: American public, 94–95; atmospheric test ban treaty, 175; Cuban Missile Crisis, 175; doing good, 258; domestic spending, 138; Hotline Agreement, 175; International Geophysical Year (IGY), 47; Kennedy (John F.) and, 179; lunar landing, 267; lunar mission, 151, 240, 265; military stalemate, 49; Moon as Cold War battlefield, 266; propaganda victories, 248; rocket plane pilots, 110; satellites, 46; space race, 10, 175, 266; thaw in, 175
Cole, Dandridge, 2–3
Collier's (magazine), 40–41, 43
Collins, Michael: Earth compared to the Moon, 241; end of space race, 252; *Gemini 10*, 198; social skills taught by NASA, 106; space walk, 198; on training, 105
Columbia University School of Journalism, 171
Columbus, Christopher: astronauts compared to, 109; commercial potential of discoveries, 256–257; financing for, 176; V-2 comparison, 17; von Braun (Wernher) and, 13
Comité Spécial de l'Année Géophysique Internationale (1957), 56
Committee on Guided Missiles, 35
Conquest of Space, The (Bonestell), 39
Conrad, Pete, 194, 198, 249
Cook, Donald, 137
Cooper, Gordon: *Faith 7* flight, 167, 173; *Gemini 5*, 194; "obit" inter-

views, 205; public face, 107; public role, 192; selection as astronaut, 105; speeches by, 172
Cord, John, 154
Corn, Joseph, 2
Cortez, Hernán, 8
Cosmonauts, 103, 114
Cost of America's space program: Apollo program, 242; domestic spending, 138, 199; lunar mission, 142–143, 151, 165, 173, 174, 188, 225; manned space flight, 81, 123, 125, 134, 137, 174, 199, 200, 242, 245; NASA budget/funding, 101–102, 135, 142, 162, 164, 169–170, 175–177, 188, 197, 200, 201, 203, 211–212, 224, 246; National Commission on Space recommendations, 261–262; satellites, 46; Senate hearings, 173–177; space exploration, 261–262
Courier-Journal (Louisville newspaper), 83
Cousins, Norman, 98–99, 268–269
Cronkite, Walter: *Apollo 11*, 235, 239; enthusiasm for space program, 171; on lunar mission, 253; NASA and the press, 106; NASA critics/opponents, 238
Crowley, John, 54
Cuban Missile Crisis, 175
Cummings, Herbert, 25
Cunningham, Walt, 192, 225–226

Dave Brubeck Quartet, 159
Declaration for Space (National Commission on Space), 261–262
Defense Intelligence Agency, 63, 100
Democracity (exhibit), 9
Democracy, lunar mission and, 237
Democratic Party: lunar mission, 143; missile gap, 91–92; presidential election (1960), 120; space race, 69–70; Sputnik, 120
DeOrsey, Leo, 112–113
Destination Moon (film), 45
Detinov, Nikolai, 31
Diamond, Edwin: *Apollo 11* media coverage, xi–xii, 238; "How to Lose the Space Race," 98; on lunar mission, 177
Dickey, James, 237
Dille, John, 107
Dingell, John, 24
Discover the Stars, 54
Discoverer 14 spy satellite, 119–120
Disney, Walt, 41
Disney Corporation, 259
Disneyland, 41–42
Dixon, George, 173
Dogs in space: Belka, 99, 120; Laika, 80*, 80–81, 84, 168, 236; Strelka, 99, 120
Domestic spending programs, 138, 199
Donovan (singer), 184
Doris Day Show (TV show), 251
Dornberger, Walter: Bell Aircraft, 182*; Hitler and, 17; on Mittelbau Dora, 21–22; rockets as weapons, 14; on V-2 rocket, 16–17; Verein für Raumschiffahrt (Society for Space Travel, VfR), 13–14; von Braun (Wernher) and, 14, 16
Douglas Aircraft, 32, 86, 137
Draper, Charles "Doc," 72
Dryden, Hugh: cosmonauts compared to astronauts, 103; Kennedy (John F.) and, 134; on lunar mission, 146; NACA, 117; NASA, 127, 191; space policy meeting, 166; speeches by, 176; technical value of manned space flight, 117
DuBridge, Lee, 85, 117

Duff, Brian: causes of public protest, 236*; on Freedom 7 flight, 140; NASA Public Affairs Division, 172; public support for NASA, 187; Webb and, 127

Duke, Charles, 235

Dulles, Allen, 57, 77

Earth, the: Collins on, 241; Moon compared to, 241; Schweickart on, 231; from space, 229–230, 241, 252; "Spaceship Earth," 229

Earth Orbit Rendezvous (EOR), 154, 155–156

Earth Satellites and the Race for Space Superiority (Stine), 68

Edwards Air Force Base, 110, 115

Ehricke, Krafft, 2–3, 11, 17

Ehrlichman, John, 244–245

Einstein, Albert, 24, 90*, 265

Eisele, Donn, 225–226

Eisenhower, Dwight: American space effort, 149; Apollo program, 123, 267–268; arms race, 46, 51; Christmas message (1958), 97–98; civilian space program, 87; farewell address, 124, 177; fiscal prudence, 143; Glennan and, 95, 123; Graham to, 74; Hornig committee report (1960), 123; ICBMs (intercontinental ballistic missiles), 52, 90–91; International Geophysical Year (IGY), 48, 50; Johnson and, 76, 212; Jupiter rocket, 52; Kennedy (John F.) and, 121, 134, 147; Little Rock, 65–66; lunar landing, 121; manned space flight, 123–124, 236–237; military-industrial complex, 87, 124; military space program, 87–88, 96; military spending, 45–46, 64, 76–77; missile gap, 91; NASA, 95; national security, 45–46, 52, 90–91;

as Neville Chamberlain, 73; nuclear weapons, 50; Open Skies initiative, 50; political fallout from Russians being first in orbit, 57; Powers (Francis Gary) and, 92; public support, 92, 143–144, 147; recruitment of astronauts, 103; reputation for failure, 118; rockets to, 59–60; Russians being first in space, 57; satellites, 46–47, 48, 51, 82; on science, 59; science advisor, 82; SCORE, 97; soldiers and scientists together, 86–87; Soviet-American relations, 50; space race, 53, 57, 81–82, 98; Sputnik, 62–63, 65–66, 76–77, 92; spying over Soviet Union, 119–120; stunts, 80, 267; Vanguard satellite, 52, 82, 87; von Braun (Wernher) and, 75, 77, 77*, 90, 95, 96–97, 101

Eisenhower administration, 67, 71, 99

Elizabeth II, Queen, 239

Ellington, Duke, 237

Enos (a chimpanzee), 157, 255

Enterprise (Starship Enterprise), 265

Environmentalists, 229, 263–264

Etzioni, Amitai: Americans' concerns about technology, 178–179; light from burning billions, 149; *The Moon-Doggle*, 177–178; NASA, 258; NASA press coverage, 171; unmanned space exploration, 258

Evening Star (newspaper), 173*

Exploration of Space (Clarke), 39

Explorer satellite, 88

Explorer II satellite, 92

Faget, Max: *Apollo 1* fire, 219–220; importance of solid-fueled launch vehicles, 54; lunar landing, 101; NASA, 191

Failures: Atlas rocket, 117–118, 119, 135, 194; Eisenhower's reputation for, 118; Kranz on, 251; *Luna 15,* 236; Mercury program, 117–118; R-7 rocket, 84, 95; Soviet failures, 84, 95, 98, 236; Vanguard satellite, 83, 87, 92; von Braun (Wernher) on, 249, 250

Faith 7 flight, 167–168, 173

Faubus, Orville, 65, 66*

Fiedermann, Angela, 19

Films: *Apollo 13,* 250; *Assignment— Outer Space,* 183; boom in space films, 183; *Breaking the Sound Barrier,* 77**; chimpanzees in space, 183; *Destination Moon,* 45; *Marooned,* 183–184; *Moon Pilot,* 183; NASA, 183–184; *Phantom Planet,* 183; *Planet of the Apes,* 183; *Star Wars,* 259–260; *When Worlds Collide,* 8; *Wild, Wild Planet,* 183

Final frontier, 42. *See also* New Frontier

Fisher, Michael, 239

Flanigan, Peter, 244

Flight into Space (Leonard), 43

Flying saucers, 8

"Footprints on the Moon" (song), 239

Ford, Gerald, 143

Ford Motor Company, 9

Forrestal, James, 35

Fort Bliss (Texas), 36

Francisco, Rio de, 159

Freedom 7 flight, 139–141, 156

Freedom of space: Sputnik, 57, 76; test of, 37, 48, 49, 51

Freeman, Fred, 40

Freud, Sigmund, 12

Friedman, Mel, 214

Friedrich, Hans, 20

Friendship 7 flight, 158–160

Fulbright, J. William, 74–75, 150–151, 247

Fuller, Buckminster, 240, 257

Fulton, James, 95, 135, 152

Futurama (exhibit), 9

G-forces, 38, 196

Gagarin, Yuri, 133–136, 159, 160

Gama, Vasco da, 109, 257

Garagiola, Joe, 237

Garnica, Mr. and Mrs. Mario, 159

Gavin, Joe, 215, 223

Gemini 4, 194

Gemini 5, 194

Gemini 6, 194

Gemini 7, 194

Gemini 8, 195–198

Gemini 9, 197–198

Gemini 10, 198

Gemini program: Agena satellite, 194, 195, 197–198; Atlas rocket, 194; complexity, 192, 198; docking, 189, 194, 195, 197–198; endurance records, 194; equipment problems, 198; extra-vehicular activity (EVA, space walking), 190, 194, 198; *Gemini 12,* 198; guided reentry, 190; Johnson and, 188; lessons from, 207; long-duration missions, 190, 194; maneuverability, 190; manned space flight, 198; pure oxygen environment, 219; regularly scheduled flights, 193, 211; required capabilities, 189–190; science, 190; sense of security, 208

General Dynamics, 103, 137, 187

General Electric, 86

General Foods, 186

General Motors, 9

German rocketry, 12–28; A-1 rocket, 15; A-2 rocket, 15; A-3 rocket, 16; A-4 rocket (*see* V-2 rocket); A-5 rocket, 16; exporting rockets from Germany to United States, 22–23;

Hitler and (*see* Hitler, Adolph); Kummersdorf, 14, 15; Mittelbau Dora, 19–22, 28; Nordhausen, 22–23; Peenemünde, 15–16, 22, 29, 30, 77; refugee German rocket scientists, 22–23, 24–27, 30, 77; slave labor, 19–22; V-2 rocket (*see* V-2 rocket)

Giancana, Sam, 153

Gilruth, Robert: lunar mission, 141, 142; manned space flight, 80; Manned Spaceflight Center (Mission Control), 169–170; NASA, 191; Shepard's planned test flight, 132; Space Task Group, 96; Spaceship NASA, 247; Webb and, 169; Weisner and, 136; Young and, 192

Glenda (a chimpanzee), 168

Glenn, John: astronaut selection procedure, 104–105; attention paid to, 109; car driven by, 115; combat missions, 107; education, 105; flight control, 163; *Friendship 7* flight, 158–160; Johnson on, 107; Kennedy (John F.) and, 114, 160; marital fidelity, 116; "obit" interviews, 205; as poster boy, 107–108; raising hands, 108; record album dedicated to, 159; religious faith, 159–160; tributes to, 159–160; on women astronauts, 168

Glenn L. Martin Company, 35, 67

Glennan, Keith: Eisenhower and, 95, 123; introduction of original astronauts, 105; on lunar mission, 147; on Mercury program, 117; NASA, 95; Project Mercury, 96; von Braun (Wernher) and, 101; Webb and, 147

Goddard, Robert Hutchins, 5–7, 12

Godwin, Francis (Domingo Gonsales), 5

Goering, Hermann, 15

Goett, Harry, 101

Goldwater, Barry, 143, 189

Gordon, Richard, 198, 249

Goya, Francisco José de, 178

GRAB (Gallactic Radiation and Background) satellite, 92

Graham, Billy, 74, 240

Grant, Ulysses S., 2

Graveline, Duane, 192

Gravity, 3, 4, 240

Gray, the, 260

Graybeal, Sidney, 144

Great Society, 189, 199

Greenglass, David, 64

Grissom, Gus: *Apollo 1*, 205–206, 208–209; Apollo Command Module simulator, 207; burial, 210; corned beef sandwich, 190–191; death, 209, 211; flight control, 163; "obit" interviews, 205; pioneer spirit, 109; public face, 107; selection as astronaut, 105; on sex, 116

Gross, H. R., 177

Group for the Investigation of Reactive Motion, 29

Gruening, Ernest, 150–151

Grumman Engineering, 187, 201, 202, 215

Guggenheim, Daniel, 6

Guggenheim, Harry, 7

Gunther, John, 64

Hagen, John, 82

Hagerty, James: announcement of IGY satellite launching, 50, 54; announcement of satellite launch, 82; Explorer satellite, 88; on refugee German rocket scientists, 77; on Sputnik, 66

"Hail to the Astronauts" (song), 184

Haise, Fred, 249–250

Hale, Edward Everett, 1–2, 5, 90

Haley, Bill, 184

Ham (a chimpanzee), 131–132, 157, 255

Hamblin, Dora, 113–114, 116

Haney, Paul, 116

Hanks, Tom, 250

"Happy Blues for John Glenn" (song), 159

Harris, Gordon, 65

Hart, Jane, 168

Heatter, Gabriel, 78

Heinlein, Robert, 45

Hesse, Herman, 267

Hewitt, Peggy, 187

High Altitude Test Vehicle (HATV), 35

High Frontier, The (O'Neill), 264

Hill, Mr. and Mrs. H. Roy, 159

Himmler, Heinrich, 16, 19, 21

Hitler, Adolph: A-3 rocket, 16; Dornberger and, 17; executions at Mittelbau Dora, 21; interest in rockets, 15, 17–18; slave labor, 19; V-2 rocket, 17; von Braun (Wernher) and, 14, 17, 18, 90

Ho Chi Minh, 230

Hodges, Eddie, 107

Holland, Spessard, 173

Holloman Air Force Base, 130

Holmes, D. Brainerd, 150

Hopkins, Sam "Lightning," 159

Hornig, Donald, 121, 245

Hornig committee report (1960), 121–123, 245

Hotline Agreement, 175

Houbolt, John, 155

Houston (Texas), 115

Houston Astros, 185

Humble Oil, 170

Humphrey, Hubert, 196–197

Hunt, Linda, 27

Huntsville (Alabama), 36, 43, 65

"Hymn of the Astronaut, The" (song), 184

I Dream of Jeannie (TV show), 183

ICBM Scientific Advisory Committee, 51–52

ICBMs (intercontinental ballistic missiles): Eisenhower and, 52, 90–91; national security, 51–52; Soviet Union, 62, 91–92; United States, 91; U. S. Air Force, 53

Inquiry into Satellite and Missile Programs (1957), 70–71

"Intergalactic Laxative" (song), 184

International Geophysical Year (IGY, 1957-1958): American plans, 50, 54; Berkner and, 47–48; broadcast frequency standard, 56; Cold War, 47; Eisenhower and, 48, 50; satellite technology, 47–51; Soviet plans, 57

Irwin, James, 241

Island One, 263–264

It's About Time (TV show), 183

Jastrow, Robert, 148, 237

Jet Propulsion Laboratory (JPL), 96

Jetsons, The (TV show), 183

"John Glenn Twist, The" (song), 159

Johnny Harris Orchestra, 239

Johnson, Lyndon: on Apollo 8, 230; on astronauts, 144; budget cuts, 199, 211–212; civil rights, 199; deadline for lunar landing, 212; domestic spending programs, 138, 199; earthrise photo, 230; Eisenhower and, 76, 212; Gemini program, 188; on Glenn, 107; Great Society, 189, 199; Kennedy (John F.) and, 170; Kerr and, 153; lunar mission, 143, 148, 152, 201,

211; Manned Spaceflight Center (Mission Control), 170, 198; missile gap, 70; Moon as a whistle stop, 145; moonlight, 150; NASA, 95, 187, 198–201; political aspirations, 98; presidential election (1964), 188–189; Smathers and, 170; on social revolution of the 1960s, 200; space policy meeting, 166; space program, 170; space race, 69–72, 120, 136–138, 138, 192, 212; Space Science Board report (1961), 137; study of American space capabilities, 136–138, 146; threats to United States from space, 69, 71, 81, 258; Vietnam War, 199, 211; Webb and, 153, 201, 212, 218–219

Johnson administration, xii

Joint Intelligence Objectives Agency (JIOA), 25, 26, 27

Joint Long Range Proving Ground, 36

Joint Research and Development Board, 35

Juno rocket, 87–88

Jupiter rocket, 52–53

"Justification books," 128

Kaczynski, Theodore (the Unabomber), 75

Kahn, Herman, 255

Kammler, Hans, 19, 20

Kapitza, Peter, 64

Kaplan, Joseph, 59

Kármán, Theodore von, 7

Karth, Joseph, 246

Kellogg, W. W., 79

Kelly, Tom, 220–221

Kennedy, Edward, 244

Kennedy, John F.: aerospace industry, 138; American growth rate, 74*; American space effort, 149–150;

assassination, 181; on astronauts, 109; Atlas rocket, 119; Bay of Pigs, 135, 147, 200; Buck Rogers and, 119; Buckley on, 177; Cold War, 179; conspiracy theorists, 182*; cooperation with Soviets in space, 121, 160–163, 162*, 173, 175, 181; deadline for lunar landing, 142, 151, 156, 165, 206–207, 208, 214–215, 220–221, 224, 232, 233, 236, 242, 248; dilemma, 180; domestic spending programs, 138; doubts about decision to go to the Moon, 179–182; Dryden and, 134; Eisenhower and, 121, 134, 147; Freedom 7 flight, 140; funeral broadcast, 182; Gagarin's flight, 133–134; Glenn and, 114, 160; grave, 181; inaugural address, 121, 142, 161; Johnson and, 170; Khrushchev and, 160; leadership, 135–136; legacy, 199; lunar mission, 141–152, 151–152, 154, 156, 165, 175, 179–182, 244; manned space flight, 128; McNamara and, 141; NAS Space Science Board report (1961), 130; NASA, 120, 127, 166–167, 179, 181, 182*; national security, 149–150, 151; New Frontier, 144–145; *New York Times* on image of, 133; political aspirations, 98; presidential campaign (1960), 119, 120, 128; prestige of the United States, 151–152; Redstone rocket, 119; research in rocketry, 72; search for space spectacular, xi; space exploration, 164; space policy meeting, 166–167; space race, 72–73, 120, 128, 135–137, 161, 164, 166–167, 181, 192; space to, 11; Sputnik, 73, 120; State of the Union speech (1961), 161; threats

to United States from space, 145;
Vietnam War, 199; von Braun
(Wernher) and, 127; Webb and,
128–129, 134, 141, 146, 166–167;
Weisner and, 124, 134, 136, 141,
152; Weisner committee report
(1961), 126–127, 130; as Winston
Churchill, 73

Kennedy, Robert, 225–226

Kennedy administration, xii

Kercheval, Jessie Lee, 202–203

Kerr, Robert, 141, 153, 160

Kerr-McGee Oil Company, 127

Kerr-McGee Oil Industries, 153

Khrushchev, Nikita: Gagarin and,
160; Gagarin's flight, 134;
Kennedy (John F.) and, 160; Korolev and, 55, 66, 189; Nixon to,
119; Soviet lunar mission plans,
179–180; Soviet per capita economic output, 71; space race, 192;
space to, 11; Sputnik, 66–67, 68; on
Vanguard satellite, 94

Killian, James, 48, 52, 82

King, Martin Luther, Jr., 224, 225

King, Stephen, 60

Kingfield, Joe, 222

Kirkpatrick, Jeanne, 262

Kistiakowsky, George, 121–122

Klaus, Samuel, 25

Kleimenov, Ivan, 29

Koch, Edward, 247

Korabl Sputnik II, 99

Korolev, Sergei: American ego, 98;
consecutive rocket firings, 163;
deputy, 189; Group for the Investigation of Reactive Motion, 29;
gulag internment, 29; Khrushchev
and, 55, 66, 189; "Preliminary Design of an Experimental World-
Circling Spaceship," 65; R-7
rocket, 55–56, 95; RD-1 rocket, 29;

rehabilitation, 29; Soviet rockets,
29, 31; von Braun (Wernher) and,
30, 55–56; Voskhod, 189, 193; on
women astronauts, 168–169

Kraft, Chris: *Apollo 1*, 209, 210; *Apollo
8*, 227; on Cunningham and Eisele,
226; deadline for lunar landing,
220

Kranz, Gene: *Apollo 1*, 207, 210; *Apollo
8*, 227; *Apollo 11*, 235; Apollo testing, 223–224; Christmas Eve message from space, 228; on failure,
251

Kummersdorf (Germany), 14, 15

Laika (a dog): flight of, 80–81; name,
80*; propaganda, 168; reentry, 84;
space race, 236

Lane, Myles, 64

Lang, Fritz, 13

Langemak, Georgiy, 29

Lapp, Ralph, 178, 242

Lawson, Alfred, 2

Leave It to Beaver (TV show), 73

Lederberg, Joshua, 175

Lenin, V. I., 5

Leonard, Jonathan Norton, 43

Leonov, Alexei: on *Apollo 1* accident,
219; on *Apollo 11* landing, 238; cosmonauts compared to astronauts,
103; end of space race, 231; *Life*
magazine coverage of astronauts,
114; space race to, 192–193; space
walk, 193; Voskhod flight, 193,
194; on women astronauts, 168

Lesher, Richard, 149

Lewis, C. S., 3–4, 229

Ley, Willy, 39

Libby, Willard, 175

Life (magazine): *Apollo 1* disaster, 211;
"astrochimp" cover, 131; *Gemini 8*,
197; Mercury astronauts, 113–115,

130; on opponents of space program, 176; Sputnik, 62; on women in space, 168

Lindbergh, Ann Morrow, 6–7

Lindbergh, Charles: astronauts compared to, 109; buses, 110; Goddard and, 6; satellite program priorities, 51–52

Lipp, J. E., 34

Little Rock (Arkansas), 65–66

Lockheed, 137

Lost in Space (TV show), 183, 195

Lovell, Bernard, 164

Lovell, James: *Apollo 8,* 227–228; *Apollo 13,* 249; *Gemini 7,* 194; Schweickart and, 231

Low, George, 101, 118–119

Luce, Clare Booth, 73, 168

Luna (spaceship), 8

Luna 15 capsule, 236

Luna Park, 8

Lunar landing, 235–240; Aldrin, xiii, 236; Aldrin's statement on setting foot on the Moon, 241; by *Apollo 12,* 249; Apollo program, 122; Armstrong (Neil) and, 235; Armstrong's statement on setting foot on the Moon, 240–241; call for (1958), 96; Cold War, 267; Communion service, 236; confidence in (1953), 41; cost, 188; deadline for, 142, 151, 156, 165, 206–207, 208, 212, 214–215, 220–221, 224, 232, 233, 236, 242, 248; Eisenhower and, 121; Faget and, 101; fuel level, 235; Hornig committee report (1960), 122–123; Mars, 261; meaning, 240–242; Mission Control, 242; NASA, 121; Nixon and, 240; predictions of, 40, 43, 90; purpose, 96; Scott on, 236; Soviet response, 238; television audience,

239; von Braun (Wernher) and, 239–240; *Washington Post* on, 230; Webb on, 261

Lunar mission, 141–152; alternative spending plan, 174; astronauts' belief in, 215; benefits from, 258; bureaucracy, 191; call for, 81; capitalism, 237; civil rights movement, 234–235; Cold War, 151, 240, 265; commercial potential of discoveries, 256–257; consensus favoring, 143–144; cost, 142–143, 151, 165, 173, 174, 188, 225; Cronkite on, 253; decision to go, 141–142, 152, 165, 175, 179–182, 200; democracy, 237; Democratic Party, 143; Diamond on, 177; Dryden and, 146; Dryden on, 146; emotion, 255–256; as entertainment, 253; finiteness, 141, 166, 242, 256, 261; frontier imagery, 150, 173; Gilruth and, 141, 142; Glennan on, 147; as historical accident, 257; hubris, 269; image management, 192; infinity of space, 145; Johnson and, 143, 148, 152, 201, 211; justification, 150, 162, 167, 180, 225; Kennedy (John F.) and, 141–152, 151–152, 154, 156, 165, 175, 179–182, 244; landing (*see* Lunar landing); last mission, 252; McNamara and, 152; meaning, 261; memorable missions, 250, 252; mountain climbing, 165; NASA, 166–167, 201, 252; national security, 150, 180; "need to explore," 257–258; opposition to, 143, 146–147, 150–151; to politicians, 192; the poor, 200; press coverage, 146; prestige of the United States, 146, 147, 150, 151–152; pub-' lic response, 238–240; public support, 141, 225; purpose, 233, 241;

repetitiveness of, 256; return on investment, 148–149; Schwarzschild on, 173; science, 174; scientific potential, 147–148, 150; Soviet plans for, 179–180; straw poll of scientists, 173–175; value to ordinary people, 256; von Braun (Wernher) and, 145, 238; Webb and, 87**, 141, 142, 166

Lunar Orbit Rendezvous (LOR), 154, 155–156

Lunar Rover, 251*

Lunarium, 5

Lunik spacecraft, 98, 100

Lunik II spacecraft, 98

Lunik III spacecraft, 98

Lunney, Glynn, 188, 203, 256

Macleish, Archibald, 230

Magellan, Ferdinand, 17, 109, 256

Mahon, George, 171, 187

Mailer, Norman, 242, 256

Mallory, George, 165

"Man in Space" (TV episode), 41–42

"Man Will Conquer Space Soon" (von Braun), 40–41

Mancini, Henry, 239

Mangan, Terence, 233

Manhattan Project, 46, 75, 86

Manned space flight: adventure, 262; aerospace industry, 137–138, 202; alternative spending plan, 174; Brick Moon, 1–2, 263; Bush (Vannevar) on, 146–147; Collier's series, 40–41; complexity, 146, 198; complexity of, 40; cost, 81, 123, 125, 134, 137, 174, 199, 200, 242, 245; docking in space, 189, 194, 195, 197–198, 231; Earth Orbit Rendezvous (EOR), 154, 155–156; Eisenhower and, 123–124, 236–237; endurance records, 194; enthusiasts for, 39–40;

first American venture, 96; first man in space, 133; first manned rocket mission, 1; Freedom 7 flight, 139–141, 156; by garment workers, 168–169; Gemini program, 198; Gilruth and, 80; Hornig committee report (1960), 121–123; inevitability of, 39–40, 146; Johnson's study of American space capabilities, 136–138; justification, xii–xiii, 123, 125, 135, 266; Kennedy (John F.) and, 128, 181; lunar mission (see Lunar mission); Lunar Orbit Rendezvous (LOR), 154, 155–156; modes of travel, 5; money, 101; NAS Space Science Board report (1961), 129–130, 137; NASA, xiv, 101–102; national security, 127; "need to explore," 257–258; personal items in space, 190–191; phasing out, 245; by pilots, 169; principles of, 1; purpose, 81, 117, 122–123, 127; reliability requirements, 123, 198; science, 129–130, 174, 242; self-perpetuation, 245–246; Slayton's defense of, 112; Soviet Union, 80–81; straw poll of scientists, 173–175; technical value, 117; threats to United States from space, 81; U. S. Army, 94; Van Allen on, 174; von Braun (Wernher) and, 38–39, 137; weightlessness, 136; Weisner committee report (1961), 124–126, 130; winged gospel, 2–4; by women, 167–169; worthiness, xii

Mansfield, Mike, 69, 246

Mariner space program, 94

Mark, Hans, 155–156

Marooned (film), 183–184

Mars: call for mission to, 81; locating, 145; lunar landing, 261; Mission to

Mars ride, 259; Paine (Tom) and, 243–247; as a state of the Union, 42; trips to, 40

Mars Project, The (von Braun), 38

Mass media, 70

Massapequa Post (newspaper), 150

Mattingly, Ken, 249

Mayflower (ship), 264

Mayo, Robert, 243, 244–245

McCormack, John, 71

McDivitt, James, 194, 206, 230–231

McDonnell Aircraft Corporation, 102–103, 127

McDonnell Douglas, 202

McElroy, Neil, 65, 65*, 65**

McKeldin, Theodore, 66*

McKinley, William, 8

McKuen, Rod, 237

McLuhan, Marshall, 237

McNamara, Robert: aerospace industry, 202; Kennedy (John F.) and, 141; lunar mission, 152; NASA, 137–138

McRoberts, Joe, 67

Mead, Margaret, 237

Medaris, John: God's will, 92–93; Jupiter rocket, 52–53; McElroy and, 65**; retirement, 101; on Russian capabilities before Sputnik, 54; soldiers in nose cones of rockets, 93; space race, 82; Sputnik, 65; Vanguard satellite, 53

Media: *Apollo 11* coverage, xi–xii, 237–238; mass media, 70; media stunts, 191, 226; response to Sputnik, 62. *See also* Films; Press coverage; Television

Medical Center (TV show), 252

Mercury astronauts, 102–120; archetype, 115; Astronaut Hall of Fame, 115; as celebrities, 112–113, 114; compensation, 114; control exer-

cised by, 110–111, 156, 163, 169; cosmonauts compared to, 103; to engineers, 111, 112; faith, 108; Gemini astronauts compared to, 192; as heroes, 117; Kennedy (John F.) on, 109; life insurance, 113; *Life* magazine, 113–115, 130; marital fidelity, 115–116; marketability, 106–109; motivations, 106; "obit" interviews, 205; pioneer spirit, 109; plans to drug them during flight, 111; popularity, 130; poster boy, 107–108; press coverage, 106, 113–115, 117, 205; psychological tests, 105; recruitment, 103–105; redundant status, 111; to rocket plane pilots, 110; selection procedure, 104–105; Slayton on, 112; social skills training, 106; space groupies, 116; as spam in a can, 110; super-fitness, 105; test pilot requirement, 103–104; wives of, 116–117; Yeager on, 110

Mercury program: astronauts (*see* Mercury astronauts); *Aurora 7* flight, 163; capsule, 102–103, 111; chimpanzees in space, 130–132, 157–158; closing of, 167; construction, 102–103; domestic spending programs, 138; door, 111; engineers, 111, 112; failures, 117–118; *Faith 7* flight, 167–168, 173; first passengers, 130; Freedom 7 flight, 139–141, 156; *Friendship 7* flight, 158–160; Glennan and, 96; hatch, 156; Hornig committee report (1960), 121–123; machine malfunctions, 158–159; pigs in space, 130; pure oxygen environment, 219; Redstone rocket, 117, 118; sense of security, 208; *Sigma 7* flight, 166; space exploration, 117; tests,

117–118, 130–133; Weisner committee report (1961), 126; windows, 111, 156

"Method of Reaching Extreme Altitudes, A" (Goddard), 5–6

Michael, Neil Edwin, 239

Michel, Jean, 21

Migliore, Daniel, 268

Military-industrial complex, 87, 124, 202

Miller, George, 171–172, 246

Mills, Wilbur, 199, 212

Mishin, Vasily, 189

Missile gap: Democratic Party, 91–92; Eisenhower and, 91; Johnson and, 70; RAND Corporation, 91; space race, 92; U. S. Air Force, 91; von Braun, (Wernher) and, 94

Missiles and Rockets (magazine), 81

Mission to Mars ride, 259

Mission to the moon. *See* Lunar mission

"Mr. Spaceman" (song), 184

Mr. Spock (fictional character), 260

Mittelbau Dora (Germany), 19–22, 28

Moley, Raymond, 146

Mondale, Walter, 215–218, 247

Moon, the: Armstrong on, 239; Anders on, 229; atmosphere, 241; bombing from, 84–85; bombing the, 79; Borman on, 229; chapel on, 233; *Clipper* flights to, 255; as Cold War battlefield, 266; cost of round trip ticket to, predicted, 255; craters on, 85; Earth compared to, 241; golfing on, 251; gravity, 240; as the high ground, 69, 258; honeymoon destination, 255; Lewis on, 229; light from, 150; lunar landing (*see* Lunar landing); lunar mission (*see* Lunar mission); Lunar Rover, 251*; military bases

on, 86–87, 87*; Moon as a whistle stop, 145; number who walked on, 255; orbiting the Moon, 227, 230–231; Scott on, 251–252; shopping on, 8; sixties zeitgeist, 266; snapshots of the dark side, 98; Sputnik, 69; as a state of the Union, 42; Teller on, 71; underground resort on, 255; uses for, 84–85; visual displays from, 79; the White House, 98

Moon-Doggle, The (Etzioni), 177–178

"Moon Maiden" (Ellington), 237

Moon Pilot (film), 183

"Moonflight" (song), 239

"Moonlight Sonata" (Beethoven), 239

Moore, Marianne, 237

Mountain climbing, 165

Movies, 8–9, 183. *See* Films

Mueller, George: *Apollo 1's* aftermath, 223–224; Apollo program testing, 232; to astronauts' wives, 221–222; corned beef sandwich, 190; migration to space, 265; NASA, 191; North American Aviation, 214; Phillips report, 216–217; Webb and, 218

Mumford, Lewis, 178, 264, 267

Murad IV, Sultan, 1

Murrary, Philip, 24

Music: album dedicated to John Glenn, 159; rock music, 184. *See also* Songs

My Favorite Martian (TV show), 183

Mylar, 185

Name That Tune (TV show), 107

NASA (National Aeronautics and Space Administration): administrators, 127, 225; Agnew and, 244–245; American way of life, 127–128; Apollo program, 102;

Army Ballistic Missile Agency (ABMA), 96, 101; black scientists and engineers, 202; Boy Scout image, 185; budget/funding, 101–102, 135, 142, 162, 164, 169–170, 175–177, 188, 197, 200, 201, 203, 211–212, 224, 246; candy machine franchise, 153; civil rights movement, 234–235; Columbia School of Journalism, 171; confidence of, 212–213, 218; Congress, 128, 142; congressional investigation of, 215–218; cost, 173, 199, 201–202; critics/opponents, 175–179, 238; deadline for lunar landing, 214–215, 233, 242; demise of, 204; Dryden and, 127, 191; Earth Orbit Rendezvous (EOR), 154, 155–156; Eisenhower and, 95; as employer, 128, 153, 203–204; entertainment industry, 170; Faget and, 191; films, 183–184; Gagarin and, 135; geographic dispersion, 206; Gilruth and, 191; Glennan and, 95; goal, 166; Great Society, 189, 199; Jet Propulsion Laboratory (JPL), 96; Johnson and, 95, 187, 198–201; "justification books," 128; Kennedy (John F.) and, 120, 127, 166–167, 179, 181, 182*; kudzu compared to, 169; leadership, 125, 153, 212–213; lunar landing, 121; lunar mission, 166–167, 201, 252; Lunar Orbit Rendezvous (LOR), 154, 155–156; managers, 191, 212–213; manned space flight, xiv, 101–102; Manned Spaceflight Center (Mission Control), 169–170, 198, 242; McDonnell Aircraft Corporation, 102–103; McNamara and, 137–138; media center, 170; media

stunts, 226; militarization of, 179; mission statement, 101–102; movie industry, 183–184; Mueller and, 191; NACA and, 95; Nixon and, 243–244, 246, 247; official song, 184; O'Hare and, 236, 236*; O'Neill and, 263; Paine (Tom) and, 243–247; Phillips report, 213–214, 216–218; politicians, coddling of, 171–173; "poor slob plan," 154; power of, 17; Powers (John) and, 96; press coverage, 171, 225, 256; priority, top, 166–167; product advertisements, 237; public affairs, 96, 172; public information officer, 234; public support, 187–188, 199–200, 246; as public works project, 202; quality control, 214–215; reputation, 218, 224; return on investment, 148–149; schedule, 211; science, 173*, 173–175, 248; Seamans and, 127, 191; self-perpetuation, 203; social scientists, 149; the South, 202–203; space exploration, 166–167; space policy meeting, 166–167; space race, 162; space-related products, 186–187; Spaceship NASA, 247; Teflon, 149; Velcro, 149; von Braun (Wernher) and, 96–97, 100–101, 260–261; wannabe astronauts, 185; *Washington Post* on, 176; Webb and, 127–129, 191; weightlessness, 190

National Academy of Sciences (NAS) Space Science Board report, 129–130, 137

National Advisory Committee for Aeronautics (NACA), 54, 95, 117

National Defense Education Act (NDEA), 75

National Geographic (magazine), 54

National Science Foundation (NSF), 75

National security: budgetary control, 45–46; Eisenhower administration, 71; Eisenhower and, 45–46, 52, 90–91; expenditures on space, 90; German rocket scientists, 24–27; ICBMs (intercontinental ballistic missiles), 51–52; Kennedy (John F.) and, 149–150, 151; lunar military bases, 86, 87*; lunar mission, 150, 180; manned space flight, 127; military spending, 45–46, 59, 76–77; moonlight, 150; Nazis, 18, 25–27; Operation Overcast, 23–24; Operation Paperclip, 24–26; satellites, 46; space, 150; space race, 179; Truman and, 45; von Braun (Wernher) and, 26, 32

National Security Council (NSC), 62, 95–96

Naval Research Laboratory (NRL), 49–50, 52

Nazism, 18, 25–27; effects of high-altitude flying, 27; Operation Overcast, 23–24; Operation Paperclip, 24–26; Rudolph and, 14, 20, 27–28; Ruff and, 27; Schreiber and, 27; security threats to United States, 24–27; slave labor building V-2s, 19–22; Strughold and, 27; von Braun (Wernher) and, 14, 15–16, 18, 22, 26

NCS-5520 (Department of Defense report), 48, 49

NCS-5522 (Department of Defense report), 49

Neal, Roy, 205

Nehru, Jawaharlal, 67

Neuberger, Maureen, 150–151

New Frontier, 144–145. *See also* Final frontier

New Priesthood, The (Lapp), 178

New Statesman (magazine), 237

New Times (magazine), 31

New York Times (newspaper): on *Apollo 11*, 239; on Apollo program, 200; on astronauts, 114; on Glenn's appearance before Congress, 159; on Goddard, 6, 12; on Kennedy's (John F.) image, 133; open letter announced in, 81; on Soviet rocketry, 56

Newell, Homer, 56, 148

Newsweek (magazine): on astronauts, 109, 114; editor, xi; "How to Lose the Space Race," 98; on press coverage of NASA, 171; on "satellite," 68; on spending on space, 246

Newton, Isaac, 7

Nikolaev, Andrian, 74**, 163

Nixon, Richard: Agnew and, 245; on *Apollo 11*, 2, 242–243; Explorer satellite, 88; to Khrushchev, 119; lunar landing, 240; NASA, 243–244, 246, 247; Paine (Tom) and, 248; political aspirations, 98; presidential campaign (1960), 119; Scott and, 231–232; space program, 231–232; space race, 120, 232; space to, 11; space travel, 246

Nordhausen (Germany), 22–23

North American Aviation: *Apollo 1*, 206, 217*; cull at, 218; government contracts, 202; Phillips report, 216; president, 217; quality of workmanship, 213–214; Webb and, 214

Nuclear energy, 10

Nuclear warfare, 32–33, 84–85

Nuclear warheads, 57

Nuclear weapons: cost effectiveness, 45; delivery systems, 31, 35; Eisenhower and, 50; Truman and, 45

Oberth, Hermann, 12–13, 15
Obi-Wan Kenobi (fictional character), 260
Obst, David, 76
Office of the Military Government United States (OMGUS), 25, 26, 27
O'Hare, Madalyn Murray, 236, 236*
O'Neill, Gerard, 2–3, 262, 263–265
Opel, Fritz von, 13
Opel-Rak 2, 13
"Open Letter to President Dwight D. Eisenhower, An" *(Missiles and Rockets)*, 81
Open Skies initiative, 50
Operation Overcast, 23–24
Operation Paperclip, 24–26
Oppenheimer, J. Robert, 71
Orbiter (previously Slug) satellite, 47, 49–50, 51
Orbitprop, 102
Order of the Ancient Astronauts, 255
Oswald, Lee Harvey, 182*
O'Toole, Thomas, 230
Out of the Silent Planet (Lewis), 3–4
Ozone layer, 32

Paine, Michael, 182*
Paine, Tom (Thomas): Abernathy and, 234–235; Agnew and, 244; *Declaration for Space,* 262–265; Flanigan and, 244; General Electric, 248; Kennedy (Edward) and, 244; *Life* magazine contract with astronauts, 115; Mars travel, 243–247; NASA, 225; National Commission on Space, 261; Nixon and, 248; prediction by, 255; Seamans and, 219; Spaceship NASA, 247–248
Pal, George, 8, 45
Pan American Exposition (1901), 8
PanAm (Pan American Airways), 255

Panofsky, Wolfgang, 149
Peale, Norman Vincent, 24
Pearson, Drew, 56, 82
Pednik, 64
Peenemünde (Germany): Soviets and, 29, 30, 77; von Braun (Werner) and, 15–16, 22
Pendray, G. Edward, 5
People (magazine), 101
Phantom Planet (film), 183
Phillips, Sam, 213–214
Pickering, William, 79
Pigs in space, 130
Planet of the Apes (film), 183
Pobeda rocket, 30
Poloskov, Sergei, 56
Poor People's Campaign, 234
"Poor slob plan," 154
Popovich, Pavel, 163
Popular Science (magazine), 9
Population explosion, 93
Poschmann, A., 19
Powers, Francis Gary, 92
Powers, John "Shorty": on critics of space program, 176; Freedom 7 flight, 140; Gagarin's flight, 133; NASA, 96
Pravda (newspaper), 62, 66
"Preliminary Design of an Experimental World-Circling Spaceship" (RAND report), 64–65
Press coverage: *Apollo 11,* 237–238; *Friendship 7* flight, 158–159; *Gemini 8,* 195; leaks to the press, 77, 97; lunar mission, 146; Mercury astronauts, 106, 113–115, 117, 205; NASA, 171, 225, 256; protocol for, 84; space race, 70. *See also entries for specific magazines and newspapers*
Project Mercury. *See* Mercury program
Proxmire, William, 199

Quarles, Donald, 48, 49, 76

R-7 rocket: failures of, 84, 95; Korolev and, 55–56, 95; launch of, 58; Redstone compared to, 55; supply of, 157
R-7 satellite: Vanguard compared to, 55
Radcliffe, Lynn, 214–215, 221
Radimer, Milt, 215
RAND Corporation: bombing the Moon, 79; founding, 32; freedom of space, 37, 48; missile gap, 91; prediction of Russian launch date, 57; "Preliminary Design of an Experimental World-Circling Spaceship," 64–65; purpose/function, 32–33; recommendations on first satellites, 33; reconnaissance satellites, 37; report on satellites (1946), 33; report on satellites (1947), 33–34
Ray, James Earl, 224
RD-1 rocket, 29
Reagan, Ronald, 255, 262
Red Star (Bogdonov), 5
Redstone Arsenal, 36
Redstone rocket: abandonment of, 157; Freedom 7 flight, 139, 140; Jupiter rocket, 52; Kennedy (John F.) and, 119; Mercury program, 117, 118, 119; R-7 compared to, 55; V-2 and, 51; von Braun (Wernher) and, 36
Reiffel, Leonard, 79
Relay 1 satellite, 182
Reston, James, 108–109, 109
Revkov, Mr. and Mrs. Alexsandr, 159
Rice University, 170
Richardson, Ralph, 77**
Rickover, Hyman, 62
Ridenour, Louis, 33
Rinehart, John, 68

Road of Tomorrow (exhibit), 9
Robots, xiii, 12
Rock music, 184
Rocket cars, 13
Rocket City (Florida), 204
Rocket into Planetary Space, The (Oberth), 12
Rocket plane pilots, 110
Rocket planes, 77
Rocket to the Moon ride, 259
Rocketdyne, 103
Rockets: Army Air Corps research budget, 32; budget for, 32; destination for, 11; Eisenhower and, 59–60; exporting rockets from Germany to United States, 22–23; first liquid-propelled rocket, 6; first test shot by von Braun, 14; funding for, 10; g-forces, 38; Goddard and, 5–6; government funding for, 10; Hitler's interest, 15, 17–18; liquid-fueled, 6, 12, 29; as logical expression of a new age, 10; missiles deployed from ships, 53; multi-stage rockets, 7; nose cones full of colored powder, 152; nuclear warheads, 57; planes distinguished from, 110; public approval, 38; rocket cars, 13; rocket planes, 77; Russian boasting, 56; soldiers in nose cones of, 93; solid-fueled launch vehicles, 54; Soviet capabilities, 30, 35; space, 5; tactical missiles, 53; thrust, 12, 94–95, 119, 122; Versailles Treaty, 13; as weapons, 7, 13–24, 30, 36
Rockets, Missiles and Space Travel (Ley), 39
"Rocking Chair on the Moon" (song), 184
Roosevelt, Theodore, 151
Rosenberg, Ethel, 64

Rosenberg, Julius, 64

Rozamus, Walter, 26

Rudolph, Arthur: Mittelbau Dora, 27–28; Nazism, 27–28; slave labor building V-2s, 19, 20; weapons-building, 14–15

Ruff, Siegfried, 27

Rumsfeld, Donald, 151

Russell, Richard, 143

Ryan, Cornelius, 40

Sagan, Carl, 79

Salinger, Pierre, 114

San Francisco Chronicle (newspaper), 175

Satellites (earth-orbiting satellites): American pre-Sputnik awareness, 67; anxiety about, 37; arms race, 51; benefits from, 258; bombing from, 69; broadcast frequency, 56; Cold War, 46; communications satellites, 182; cost effectiveness, 46; docking with, 195; Eisenhower and, 46–47, 48, 51, 82; enthusiasm for developing, 35; equatorial orbits, 37; feasibility studies, 35; first American launch attempt, 82–84; first deployed by United States, 88; freedom of space, 37, 48, 49, 51; human imagination, 33; Inquiry into Satellite and Missile Programs (1957), 70–71; International Geophysical Year (IGY), 47–51; interplanetary travel, 33; legality of spying from space, 37, 48; national security, 46; navigation satellites, 99; navy report, 35; nuclear warfare, 32–33; Open Skies initiative, 50; permission for Army to launch, 82; political fallout from Russians being first in orbit, 57; prestige from being first, 49; RAND report

(1946), 33; RAND report (1947), 33–34; reconnaissance satellites, 37, 48–49, 50; satellite program priorities, 51–52; scientific satellites, 48–50; spy planes, 92; spy satellites, 92, 119–120; surveillance satellites, 258; as symbols of cultural virility, 33; usefulness of, 37–38; on Vanguard satellite, 93–94; von Braun (Wernher) and, 47; weather satellites, 99, 148

Saturday Evening Post (magazine), 9, 267

Saturn rocket: Army Ordnance Missile Command study, 86; improvements to, 122; speed, 265; von Braun (Wernher) and, 100–101

Saturn V rocket: Apollo 11, 235; Rudolph and, 27; stages, 224; von Braun (Wernher) and, 227

Scheer, Julian, 184, 234

Schirra, Walter: Apollo 7, 225–226; car driven by, 115; corned beef sandwich, 190; faith, 108; Gemini 6, 194; "obit" interviews, 205; public face, 107; selection as astronaut, 105; Sigma 7 flight, 166

Schomer, Howard, 93

Schreiber, Walter, 27

Schriever, Bernard, 87*

Schultze, Charles, 200–201, 211–212, 252–253

Schwarzschild, Martin, 173

Schweickart, Rusty, 206, 230–231

Science: Aurora 7 flight, 163; black scientists, 202; diversion of scientists to space program, 174; Eisenhower on, 59; Gemini program, 190; great societies, 87; lunar mission, 147–148, 150, 174; manned space flight, 129–130, 174, 242; military-industrial complex, 87;

NASA, 173*, 173–175, 248; politics, 175; R and D, value of, 87; soldiers and, 86–87; space travel, 191; Sputnik, 63, 77–78; state-sponsored research, 59; symbols of progress, 146; Weisner committee report (1961), 125

Science fiction movies, 8–9

Science (magazine), 173–174

SCORE (Signal Communications by Orbiting Relay Equipment), 97–98

Scott, David: *Apollo 1*, 206; *Apollo 9*, 230–232; *Apollo 15*, 251; Apollo deadline, 206–207; complexity of Gemini missions, 192; *Gemini 8*, 195–197; on Kennedy (John F.), 182; lunar landing, 236; on the Moon, 251–252; NASA budget, 175–176; NASA's selection procedures, 106; Nixon and, 231–232; space program's value, 251

Scott, Lurton, 196

Scott, Robert, 109

Seale, Leonard, 154

Seamans, Robert: *Apollo 1*, 209, 224; congressional investigation of NASA, 217–218; *Gemini 8*, 196–197; NASA, 127, 191; Paine (Tom) and, 219; Phillips report, 216, 218; space policy meeting, 166; Webb and, 200, 216, 217–218, 219

Sedov, Leonid, 54, 74, 162

See, Elliott, 197

Seinna, Sister Mary, 158*

Sevareid, Eric, 241–242, 256

Shea, Joe, 207–208, 218

Shell Oil Company, 184

Shepard, Alan: "A-OK," 140; *Apollo 1*, 223; *Apollo 8*, 227; *Apollo 14*, 251; on chimpanzees in space, 131; flight control, 163; Freedom 7

flight, 139–141, 156; golfing on the Moon, 251; "obit" interviews, 205; pioneer spirit, 109; planned test flight, 132–133; public face, 107; selection as astronaut, 105; space flight, 139–141; tributes to, 159

Sherr, Lynn, 171

Sigma 7 flight, 166

Silverstein, Abe, 102, 118

Skurla, George, 220, 221

Slayton, Deke: *Apollo 8*, 227; on astronauts, 112; heart anomaly, 163; public face, 107; selection as astronaut, 105; unmanned space flight, 111; Yeager and, 112

Slug (later Orbiter) satellite, 47, 49–50, 51

Smathers, George, 170

Smith, Bernard, 4

Smith, Margaret Chase, 246

Society of Experimental Test Pilots, 112

Sockman, Ralph, 133–134

Songs: "The Ballad of John Glenn," 159; "Bless Thou the Astronauts," 184; "Footprints on the Moon," 239; "Hail to the Astronauts," 184; "Happy Blues for John Glenn," 159; "The Hymn of the Astronaut," 184; "Intergalactic Laxative," 184; "The John Glenn Twist," 159; "Mr. Spaceman," 184; "Moonflight," 239; NASA official song, 184; rock music, 184; "Rocking Chair on the Moon," 184; "Space Oddity," 184; "Spaceman Twist," 159; "Telstar," 184, 184*; "The Walk of Ed White," 194

South, the, 202–203

Soviet Union: American view of, 64; atomic blast (1949), 35, 37; atomic weapons, 55; boasting about rockets, 56; Bush (Vannevar) on, 34; capabilities of, 35, 62, 64–65, 68–69, 81, 162; education in, 76; growth rate, 74; ICBMs (intercontinental ballistic missiles), 62, 91–92; International Geophysical Year (IGY), 57; long-duration missions, 190; manned space flight, 80–81; missile sites, 91; offer of financial aid to United States, 83; offer to help find *Aurora 7*, 163; Peenemünde, 29, 30, 77; per capita economic output, 71; Pobeda rocket, 30; prestige, 67; refugee German rocket scientists, 30, 77; rendezvous in space, 167; rocket thrust, 94–95, 119; space program, 30; space race, 179–180, 238; spying over, 119–120; state-sponsored scientific research, 59; threats from space to, 84–85; usefulness of satellites, 37–38; V-2 rockets in, 30; "white envy," 238

Soyuz capsule, 193

Soyuz program, 189

Space: as adventure, 11; American Dream, 9; Americans' interest in, 7–10; Christmas Eve message from, 227, 228; docking in, 189, 194, 195, 197–198, 231; to dreamers, 11; extra-vehicular activity (EVA, space walking), 190, 193, 194, 198; films (*see* Films); freedom of (*see* Freedom of space); human imagination, 4–5; infinity of, 145; man's limits, 266; migration to, 265; movies, 183; national security, 150; pioneer spirit, 109; to pragmatists, 11; romance of, 182; space-related products, 185–187; spying from, 37, 48, 92, 119–120; television (*see* Television); threats to Soviet Union from, 84–85; threats to United States from, 41, 45, 62, 68–69, 71, 81, 88–89, 145; toys, 184–185

Space age, origins of, 21

Space Age Management (Webb), 258–259

Space Education Club, 184

Space exploration: Armstrong and, 257; cost, 261–262; Etzioni and, 258; funding for, xiii; Kennedy (John F.) and, 164; Mercury program, 117; NASA, 166–167; space race, 129; Truman administration, 32; Tsiolkovsky and, 152, 262–263; Weisner and, 261

Space Food Sticks, 186

Space groupies, 116, 221

"Space Oddity" (song), 184

Space race: American complacency about, 7, 34, 35, 40–41, 54, 56, 68, 73, 74, 83, 100, 119, 150; American lead, 99*, 120, 125, 135, 156–157, 162, 182, 193; Asimov on, 261; astronauts, 192–193; chimpanzees in space (*see* Chimpanzees in space); Christmas Eve message from space, 227, 228; Cold War, 10, 175, 266; college football compared to, 144; commodification of, 186–187; congressional support, 95; consensus favoring, 259; construction plants workers, 103; cooperation with Soviets in space, 121, 160–163, 162*, 173, 175, 181; as cure for America's woes, 78; Democratic Party, 69–70; dogs in space (*see* Dogs in space); Eisenhower and, 53, 57, 81–82, 98; end of, 231, 242, 252; first American to orbit the Earth, 157; first man in space, 133; first manually controlled re-entry, 193; first satellite deployed by United States, 88; first space walk, 193; frontier imagery, 129, 144–145; funding for, xiii; ground rules, xii; Hornig committee report (1960), 122; inevitability of, 34; inter-service rivalries, 81; Johnson and, 69–72, 120, 136–138, 138, 192, 212; Kennedy (John F.) and, 72–73, 120, 128, 135–137, 161, 164, 166–167, 181, 192; Khrushchev and, 192; Laika, 236; liberal democracy, 10–11; as lifting contest, 94–95; long-duration missions, 190, 194; lunar mission, 180; magnificence of, 193; mass media, 70; meaning, 233; Medaris and, 82; military-industrial complex, 87; missile gap, 92; money, 101–102; NASA, 162; national security, 179; Nixon and, 120, 232; objectives, 87; to politicians, 192; press coverage, 70; purpose, 88, 92, 122, 266; raison d' être, 266; refugee German rocket scientists, 22–23, 24–27, 30, 77; rendezvous in space, 164, 167; shallowness of, 99; slogans in, 70; Soviet Union, 179–180, 238; space exploration, 129; space policy meeting, 166–167; Sputnik, 67; Teller on, 71; timetable for space exploits, 95–96; von Braun (Wernher) and, 34, 82; Weisner committee report (1961), 125; women in space, 167–169

Space Science Board report (1961), 129–130, 137

Space Shuttle: adventure, 262; expectations, 265; performance, xii; present era, xiv

Space stations: Earth Orbit Rendezvous (EOR), 155–156; Oberth and, 12; prerequisite to building, 164; threats to United States from, 41, 45; von Braun (Wernher) and, 40

Space Task Group, 96, 245

Space travel: adventure, 262; American belief in, 43; *Apollo 8,* 229; chimpanzees in space (*see* Chimpanzees in space); *Clipper* flights to the Moon, 255; cost of round trip ticket to the Moon, predicted, 255; discovery of a perfect world, 268; dogs in space (*see* Dogs in space); dreams of overcoming gravity, 4; ego tripping, 146; fantasies of, 92–93, 259–266; frontier imagery, 42; Fuller on, 257; God's will, 93; interplanetary travel, 243–247; justification, 266; Moon as a whistle stop, 145; "need to explore," 257–258; Nixon and, 246; pigs in space, 130; population explosion, 93; purpose, 42–43, 71, 117; robots, xiii; rock music, 184; romance, 191; science, 191; Sputnik, 69; vicarious enjoyment of, 186; von Braun (Wernher) and, 14, 38–39, 42–43, 93; women in space, 167–169. *See also* Manned space flight

Space walks: Cernan, 198; Collins, 198; *Gemini 9,* 190; Gemini program, 190, 194, 198; Leonov and, 193; McDivitt, 194; White, 194, 198

"Spaceman Twist" (song), 159

"Spaceship Earth," 229

Spaceship NASA, 247–248

Spam in a can, 110

Speer, Albert, 17, 18, 19–20

Sputnik, 54–78; in American English, 63; American prestige, 67; American reaction, 62; American response, 67–78; arms race, 68;

beeps from, 63; CIA response, 62, 63; Clarke on, 63; Defense Intelligence Agency response, 63; Democratic Party, 120; Egyptian response, 67; Eisenhower and, 62–63, 65–66, 76–77, 92; Faget's question, 54; fall from orbit, 78; foods based on, 64; freedom of space, 57, 76; Gagarin and, 135; Hagerty on, 66; Indian response, 67; Kennedy (John F.) and, 73, 120; Khrushchev and, 66–67, 68; launch of, 58; launch rocket, 55–56; *Leave It to Beaver*, 73; *Life* on, 62; Little Rock, 66; Medaris and, 65; media response, 62; military spending, 59; the Moon, 69; NSC response, 62; panic, 68–69; Pearl Harbor compared to, 76, 78; predictions of, 56–57; presidential election (1960), 120; products based on, 64; rocket planes compared to, 77–78; science behind, 63, 77–78; shock of, 267; sound, 63; Soviet prestige, 67; space race, 67; space travel, 69; stock market response, 63–64; as stolen American technology, 64–65; Teller on, 68–69; UFO sightings, 260; Vanguard compared to, 94; von Braun (Wernher) and, 65; weight, 63; Yeager on, 77–78
Sputnik II, 80, 84
Sputnik III, 94, 97
Spying: legality of spying from space, 37, 48; spy planes, 92; spy satellites, 92, 119–120
"Squawk boxes," 195
Stafford, Tom, 194, 197–198, 232
Stalin, Joseph: ballistic missiles, 31, 35; death, 55; Korolev and, 29; on Truman, 31

Stanyukovich, Kirill, 69
Star Trek (TV show), 183, 259–260
Star Wars (film), 259–260
Steenbeke, René, 20, 21
Stehling, Kurt, 83
Stennis, John, 150
Stewart, Homer, 49, 50
Stine, G. Harry, 68
Strauss, Lewis, 10
Strelka (a dog), 99, 120
Strughold, Hubertus, 27
Stuhlinger, Ernst, 54
Surveyor 3 lunar probe, 249
Swigert, Jack, 249

Tang, 185–186
Taylor, Maxwell, 86
Teague, Olin, 150, 246
Technological Capabilities Panel, 48
Technology, Americans' concerns about, 178–179
Teflon, 149, 185
Television: *All in the Family*, 251; audience for lunar landing, xi–xii, 239; *Doris Day Show*, 251; *I Dream of Jeannie*, 183; *It's About Time*, 183; *The Jetsons*, 183; *Leave It to Beaver*, 73; *Lost in Space*, 183, 195; *Medical Center*, 252; *My Favorite Martian*, 183; *Name That Tune*, 107; *Star Trek*, 183, 259–260; *Tom Corbett: Space Cadet*, 9; *Tonight Show*, 252
Teller, Edward: the Moon, 71; power, 124; on rocket scientists, 164; space race, 71; on Sputnik, 68–69
"Telstar" (song), 184, 184*
Tereshkova, Valentina, 74**, 167–169
Theology Today (journal), 268
Thomas, Albert, 170, 177
Thompson, Frederick, 8
Three Laws of Astronautics, 3
Time in Outer Space (album), 159

Time (magazine): on *Collier's* series, 41; Men of the Year (1968), 228; on NASA under Webb, 128; picture of astronauts, 176

Tiros 1 satellite, 99

Titan II rocket, 194

Titov, Gherman, 133, 156, 157

Tokady, Grigory, 29, 30

Tom Corbett: Space Cadet (TV show), 9

Tonight Show (TV show), 252

Tornadoes, The, 184*

Toys, 184–185

Transit 1B satellite, 99

Trip to the Moon, A (exhibit), 8

Truman, Harry S.: Bush (Vannevar) and, 34; national security, 45; nuclear weapons, 45; Operation Overcast, 24; Operation Paperclip, 24–26; Stalin on, 31; Webb and, 127

Truman administration, 32

Tryon, Tom, 183

Tsiolkovsky, Konstantin: centennial of his birth, 56; "Father of Cosmonautics," 5; Goddard compared to, 5, 6; pull of space, 4–5; space exploration, 152, 262–263; space to, 11; Verne and, 4; von Braun (Wernher) and, 90

Turner, Frederick Jackson, 42

Turnill, Reginald, 224–225

TV Guide (magazine), 239

UFO sightings, 260

United States: bankruptcy, 267; consensus in, 259; domestic spending programs, 138; education in, 75–76; growth rate, 74; ICBMs (intercontinental ballistic missiles), 91; love affair with science and technology, 267; naiveté, 117; Operation Overcast, 23–24; presidential campaign (1960), 119, 120;

prestige, 67, 99, 124–125, 134, 141–142, 146, 147, 150, 151–152; reaction to Sputnik, 62; refugee German rocket scientists, 22–23, 24–27, 30, 77; second-strike capability, 84–85; Soviet offer of financial aid to, 83; space program, 30; state-sponsored scientific research, 59; threats from space to, 41, 45, 62, 68–69, 71, 81, 88–89, 145; usefulness of satellites, 37–38; V-2 rockets in, 22–23, 31–32, 36–37; worth of, 99

U. S. Air Force: budget, 91; Edwards Air Force Base, 110, 115; Holloman Air Force Base, 130; ICBMs (intercontinental ballistic missiles), 53; Jupiter rocket, 53; lunar bases, 86; missile gap, 91

U. S. Army: Air Corps rocket research budget, 32; Army Ballistic Missile Agency (ABMA), 49, 53, 55, 96; Army Ordnance Missile Command, 86; Jet Propulsion Laboratory (JPL), 96; Jupiter rocket, 53; manned space flight, 94; Orbiter (previously Slug) satellite, 47, 49–50, 51; permission to launch satellites, 82; platoon-in-a-can plan, 93; requirements for space transportation and combat, 85; tactical missiles, 53; von Braun (Wernher) and, 96

U. S. Department of Defense, 35, 100

U. S. Information Agency (USIA), 67

U. S. Navy: High Altitude Test Vehicle (HATV), 35; missiles deployed from ships, 53; report on satellite technology, 53; Vanguard satellite, 49–50, 52, 53, 55, 67; Viking rocket, 49–50, 54

U. S. State Department, 25–26

U. S. War Department, 24–25, 35
Up With People, 194
Urey, Harold, 175, 237

V-2 rocket: Albert rockets, 38; Columbus compared to, 17; Dornberger on, 16–17; Hitler and, 17; Juno rocket, 88; name, 16*; Nazi Germany, 16–22; ozone layer, 32; Redstone rocket, 51; slave labor, 19–22; in Soviet Union, 30; Tomorrowland, 42; in United States, 22–23, 31–32, 36–37
Valier, Max, 13
Van Allen, James, 47, 174
Vanguard. *See* Viking/Vanguard
Velcro, 149, 185, 209
Ventures, The, 184
Venus, 81, 146, 261
Venus, Vik, 239
Verein für Raumschiffahrt (Society for Space Travel, VfR), 13–14, 29
Verne, Jules: demand for works by, 61; Goddard compared to, 6; means of space travel, 5; Oberth and, 12; Tsiolkovsky and, 4
Versailles, Treaty of (1919), 13
Vietnam War, 199, 211, 230
Viking/Vanguard: builder, 67; confidence in, 52, 53, 82; Eisenhower and, 52, 82, 87; failures, 83, 87, 92; Khrushchev on, 94; Medaris and, 53; Naval Research Laboratory, 49–50; as precursor, 94; progress on, 55; R-7 compared to, 55; science, 93–94; Sputnik compared to, 94; successes, 93; supplementary program, 82; thrust, 235; TV-3 test, 82–84; TV-5 test, 82; U. S. Navy, 54; Van Allen and, 174; von Braun (Wernher) and, 52, 77
Viorst, Milton, 153

Von Braun, Magnus, 21, 22, 52–53
Von Braun, Wernher: A-1 rocket, 15; A-2 rocket, 15; A-3 rocket, 16; A-5 rocket, 16; ambitions, 38–39; American people, 44; on American science, 75; Apollo program legacy, 260–261; Army Ballistic Missile Agency (ABMA), 53; artificial star, 43; at Cape Canaveral, 36; Castenholz on, 90; citizenship, 43; Clarke and, 53; in *Collier's*, 40–41; Columbus and, 13; commencement address, 43–44; death, 254; desist order, 214; Dornberger and, 14, 16; egotism, xii–xiii; Eisenhower and, 75, 77, 77*, 88, 90, 95, 96–97, 101; entrance visa to United States, 28; exporting rockets from Germany to United States, 22–23; on failure, 249, 250; fear mongering, 88–90; first test shot, 14; at Fort Bliss, 36; Gestapo, arrest by, 18; Glennan and, 101; Hale and, 90; as helicopter salesman, 254; Himmler and, 16; Hitler and, 14, 17, 18, 90; at Huntsville, 36, 43, 65; Juno rocket, 87–89; Kennedy (John F.) and, 127; Korolev and, 30, 55–56; at Kummersdorf, 14, 15; lunar landing, 90, 239–240; lunar mission, 145, 238; "Man in Space" (TV episode), 41; "Man Will Conquer Space Soon," 40–41; manned space flight, 137; marriage, 28; McElroy and, 65; mind of, 23; missile gap, 94; at Mittelbau Dora, 21; NASA, 96–97, 100–101, 260–261; national security, 26, 32; Nazism, 14, 15–16, 18, 22, 26; Oberth and, 15; on Operation Overcast, 24; optimism of, 41; peacetime equivalent of total war, 89–90; at Peen-

emünde, 15–16, 22, 36; population explosion, 93; power, 124; predictions by, 90; press leak by, 77, 97; as prophet, 75; publicity, 13; Redstone rocket, 36; rendezvous in space, 155; Rocket to the Moon ride, 259; satellites, 47; Saturn rocket, 100–101; Saturn V rocket, 227; science fiction novels, 13; slave labor building V-2s, 19, 21; Slug (later Orbiter) satellite, 47; smiling, 253–254; soldiers in nose cones of rockets, 93; Soviet atomic blast (1949), 35–36; Soviet intentions, 89; space race, 34, 82; space stations, 40; space travel, 14, 38–39, 42–43, 93; Spaceship NASA, 247; Speer and, 18; as spin doctor for space, 90; Sputnik, 65; SS service, 16, 26; on technology and ethics, 43–44; threats to United States from space, 88–89; Tomorrowland, 41; Truman administration, 32; truth to, 14; Tsiolkovsky and, 90; understanding of Americans, xii–xiii; U. S. Army, 96; Vanguard satellite, 52, 77; Weisner and, 132; on women astronauts, 168

Von Puttkamer, Jesco, 155–156, 254

Voorhest, Bill, 222

Voskhod capsule, 189, 193, 194

Vostok 2 rocket, 156

Vostok 3 rocket, 163–164

Vostok 4 rocket, 163–164

Vostok capsule, 133

Voyager space probes, 94

"Walk of Ed White, The" (song), 194

"Walking Billy Blastoff" (toy), 184

Wall Street Journal (newspaper), 68, 215

Walsh, J. Paul, 82–83

Walters, Barbara, 237

War of the Worlds, The (radio broadcast), 8

Washington Post (newspaper): Kennedy on lunar mission, 151–152; on lunar landing, 230; on NASA, 176; Space Food Sticks ad, 186

Washington Star (newspaper), 184

Weaver, James, 151

Weaver, Warren, 174

Webb, James: "administrator's discount," 142; ambition, 128; benefits from NASA, 206; Berkner and, 129, 130; congressional investigation of NASA, 216–218; congressional support for NASA, 172; on control of space, 176–177; Duff and, 127; frontier imagery, 144–145; Fulton and, 135; Gilruth and, 169; Glennan and, 147; Johnson and, 153, 201, 212, 218–219; Kennedy (John F.) and, 128–129, 134, 141, 146, 166–167; Kerr-McGee Oil Company, 127; likelihood of space tragedy, 208; lunar landing, 261; lunar mission, 87**, 141, 142, 166; management, 200, 212–213; McDonnell Aircraft Corporation, 127; Mondale and, 218; Mueller and, 218; NASA, 127–129, 191; North American Aviation, 214; outside investigations of NASA, 215; Phillips report, 216–218; retirement from NASA, 218–219; Seamans and, 200, 216, 217–218; *Space Age Management*, 258–259; space policy meeting, 166–167; speeches by, 172, 176; Truman and, 127; weather satellites, 148; Weisner and, 166

Weightlessness, 3, 136, 190

Weisner, Jerome: aerospace industry, 138; cost of space exploration, 261; Kennedy (John F.) and, 124, 128, 134, 136, 141, 152; NASA budget, 135; space policy meeting, 166; von Braun (Wernher) and, 132; Webb and, 166; weightlessness, 136; Weisner committee report (1961), 124–127, 130

Weisner committee report (1961), 124–125, 130, 245

Welles, Orson, 8, 237

Wells, H. G., 12

West, Roy, 159

Wev, Bosquet, 25–27

Wheelon, Albert, 144

When Worlds Collide (film), 8

White, Ed (Edward): *Apollo 1,* 205–206, 208–209; burial, 210; death, 209, 211; *Gemini 4,* 194; space walk, 194, 198; "The Walk of Ed White," 194; *Time* magazine, 176

Whitehead, Clay, 243–244

Whittaker, Walter, 171

Wickersham, Victor, 147

Wicky (reporter), 116

Wild, Wild Planet (film), 183

Williams, John, 61

Williams, Mennen, 66

Wilson, Charles, 63

Wilson, Glen, 137

Winged gospel, 2–4

Wise, Stephen, 24

Wolfe, Tom, 83

Women: astronauts' wives, 113, 116–117, 195–196, 221–222; Tereshkova's flight, 167–169

World's Fair (1939, New York), 9

World's Fair (1958, Brussels), 100

World's Fair (1962, Seattle), 160

X-1 rocket plane, 77

X-20 (Dyna-Soar) rocket plane, 78

Yeager, Chuck: on astronauts, 110; *Declaration for Space,* 262; Slayton and, 112; on Sputnik, 77–78

York, Herbert, 91

Young, John: affair, 192; *Apollo 10,* 232; corned beef sandwich, 190; *Gemini 10,* 198; Gilruth and, 192

Young, Whitney, 200

Zond 7 capsule, 227

About the Author

Gerard J. DeGroot is Professor of Modern History at the University of St. Andrews in Scotland. His many books include *The Bomb: A Life*, *The First World War*, and *A Noble Cause? America and the Vietnam War*.